PLANNING RESEARCH

PLANNING RESEARCH

A CONCISE GUIDE FOR
THE ENVIRONMENTAL AND
NATURAL RESOURCE SCIENCES

John C. Gordon

Yale University Press / New Haven and London

The following sources have been excerpted with permission: D. E. Koshland, "Two Plus Two Equals Five" (Editorial), *Science* 247, no. 4949 (1990): 1381, Copyright © 1990 AAAS; and T. S. Kuhn, *The Structure of Scientific Revolutions,* 2nd ed. (Chicago: University of Chicago Press, 1970), Copyright © 1962, 1970 by the University of Chicago.

Published with assistance from the Louis Stern Memorial Fund.

Designed by Mary Valencia.

Set in Minion and Gill Sans by The Composing Room of Michigan, Inc.

Printed in the United States of America by Vail-Ballou Press, Binghamton, New York.

Library of Congress Cataloging-in-Publication Data

Gordon, J. C. (John C.), 1939–

　　Planning research : a concise guide for the environmental and natural resource sciences / John C. Gordon.

　　　　p. cm.

　　Includes bibliographical references and index.

　　ISBN: 978-0-300-12007-3 (clothbound : alk. paper)

　　ISBN: 978-0-300-12006-6 (pbk. : alk. paper)

　　1. Environmental sciences—Research.　2. Environmental sciences—Experiments.

I. Title.

　　GE70.G695 2007

　　363.7'0072—dc22

2006026960

A catalogue record for this book is available from the British Library.

The paper in this book meets the guidelines for permanence and durability of the Committee on Production Guidelines for Book Longevity of the Council on Library Resources.

10 9 8 7 6 5 4 3 2 1

CONTENTS

PREFACE

This book is the result of my taking a course in writing study plans from Carl Stoltenberg in 1964, and then trying to teach such a course at three different universities for more than thirty years. It is intended to be a primer for those who haven't done much, or any, scientific research. The focus is on thinking carefully and writing a plan before starting to attack the research in the laboratory or field. I have learned continually from the students in my course and from the various people who have helped me teach it. I dedicate the book to all of them, and particularly to Carl Stoltenberg, John Beuter, and Bill Farrell, all mentors and colleagues.

ONE

The Importance of Written Plans

In the current science environment, money and permission to do a specific piece of research depend absolutely on the quality of a written document explaining what research is to be done and why. This is true regardless of the discipline, location, or size of a scientific project. It has been fashionable to debate whether written plans for research are simply bureaucratic impediments to real science. For some exceptional individuals this may be true. For most, however, two compelling reasons for doing written plans remain. First, the system requires them, and, more important, thinking, especially thinking before doing, is the key to good research.

WHY WRITE STUDY PLANS?

Science, as a rigorous form of honesty, is a difficult pursuit. We like to be right, but science proceeds only by proving things wrong. To quote my doctoral adviser, Walter Loomis (personal communication), "Science is what can be proven wrong (if it is wrong); philosophy is what you can only argue about." Good science usually demands long preparation, repetition of difficult or boring measurements over long periods of time, ruthless discard of failed ideas, and an exhausting struggle to bring results to publication. Perhaps the most daunting aspect of a scientific career is that most or all of it usually will be taken up doing what Thomas Kuhn (1970) called "normal science"; that is, producing those small advances that together eventually bring important new understanding. The giants of science are few, but those who contribute research to the creation of giant insights are very many.

The quality of the research done by the many is a critical determinant of the rate of scientific advance and of the quality of the eventual giant insights. Quality control over scientific research is maintained through careful planning and review of proposed research efforts. Researchers thus use written study plans to carefully describe what they propose doing and, particularly, why. In the absence of a written plan, determining the "why" is particularly difficult. But, as Gordon's second law of science says, a research project not worth doing at all, is not worth doing well.* Only by explaining the "why" of a research project can the skeptical be convinced that a study is worth doing. This is especially important for research that is to be the basis for a dissertation or a thesis. Often, the thesis problem is a piece of a larger problem hierarchy that addresses an aspect of theory or practice of great importance. The learning researcher must be able to demonstrate how his or her study links up to this larger research question, the question that will specify the observations and theories that frame and underpin the thesis problem. The researcher will use these observations and theories to generate the study's research objective and hypotheses (see Chapters 4 and 5). This thought process is also necessary to make the student more than a "pair of hands" and a full participant in the larger world of science.

Each piece of the larger problem is subject to a progression from ab-

---

* "Gordon's second law of science" was in common usage during the 1960s, probably taken from *Worm Runners Journal*. Independently formulated by the author in Carl Stoltenberg's research methods class at Iowa State University in 1964.

stract idea to concrete, written research product. In this progression, the initial idea is transformed into a written study plan, for the reasons given below. The study plan then metamorphoses over the course of the research into written manuscript to be judged by the researcher's peers. These "pieces," as they are produced and judged sound, are continually woven into the ever-changing whole that is the answer to the overarching question and its place in the theoretical and operational body of science.

Those engaged in research on environmental and natural resource topics have an additional burden. Humans have a tendency to draw a boundary between "us" and "the natural world," and this greatly affects how people think about "the environment." For the environmental researcher it means that there is a double uncertainty principle. The researcher has, of course, to confront her own potential influence on the system to be studied, which may be a key, or rare, component of the environment itself. But the environmental scientist must also be specifically aware of the probable influence of humanity at large. All earthly natural systems are probably affected by human activity either on purpose or in collateral ways.

Environmental problems, issues that arise out of concern for environmental values and are thus often the object of environmental research, are especially difficult to solve. Environmental and natural resource sciences are quite young, and the problems they address tend to have a weak and scattered science base, take long times to solve, and involve a daunting array of disciplines. They also are frequently the subject of controversy and conflicting opinions based on differing human values (Gordon and Berry 1993). Although the basic methods of scientific thought and action are reasonably universal across problems and disciplines, planning environmental research often requires special emphasis on problem selection and definition, setting specific and achievable objectives, and planning for long-term, logistically difficult field activities. Thus, these topics receive special emphasis here.

There are many reasons for doing written study plans. All of them apply with special force to environmental research. Writing a plan helps you to do the following important things:

1. *Think as carefully and thoroughly as possible before doing.* Science resources are scarce. Most research projects are expensive in human time, equipment, and supplies. Research funders expect these resources to be spent wisely and to find reviewed plans reassuring.

2. *Get help.* A written plan can be reviewed by people outside your immediate environment. With e-mail, reviewers can come from virtually anywhere. Few research projects are so well conceived initially that they can't be improved with expert advice.

3. *Get money.* Most research is funded on the basis of written plans, or "proposals." The ability to write clear plans is the most important skill to use to get money to do research.

4. *Provide for continuity.* Your research is important or it wouldn't be happening. Thus, it is important to carry it through to a usable result, if at all possible. Many environmental and natural resource research projects take place over extremely long periods of time. The originating scientist may not be the one that finishes. A sound written plan will ensure a useful ending. But continuity is important even if the same researcher stays the course. Inevitably, minds and methods change during the course of a research effort. A written plan will reveal what changed, and when, and what the changes mean from the retrospective view from the results. It is important to read your plan periodically and note changes in writing. One technique is to use a "track changes" editing program on a word processor and ask the initial reviewers to look over changes periodically during the course of the research.

5. *Provide for coordination.* Written plans provide the means for making sure research projects on similar topics and problems fit together as well as possible. If written plans are created and widely shared, they also could lead to interdisciplinary research on problems that don't seem to fit together at all, at first. Shared hypotheses and methods could be made to fit together better and to lead more directly to giant insights.

THE INTENDED AUDIENCE AND GOAL

This book is the result of a course in research methods that I have taught for thirty years to graduate students in three different universities. This book is intended for students who are at the beginning of a research career or for those who will use research results in professional practice. It may also be useful to established researchers who want a refresher course on research planning and its underlying philosophy.

This book will help you to:

- Prepare complete study plans and grant proposals;
- Explain several views of how and why science works;
- Recognize the major conceptual limitations of scientific research;
- Constructively criticize study plans;
- Communicate research plans to specific audiences;
- Outline the structure of environmental and natural resource research;
- Think creatively about science as a human activity.

The core method used to approach this ambitious list is the stepwise preparation of a written study plan.

USING THIS BOOK

This book is intended as a text to be used with the cited readings for a one-semester graduate course on research planning for students in environmental and natural resource curricula. The goal of the book is to give practical help in planning and doing research, and in using its results. In the course, students prepare a real study plan in addition to the presentation, reading, and discussion of the text and other assigned readings. The plan is prepared in pieces, with each piece reviewed by the instructor, and each student presents a plan orally to the class at the end of the term. If you are reading the book outside the discipline of a scheduled class, you will probably gain the most from it if you attempt to prepare a plan while reading the book. It also helps to discuss the plan pieces with others as you create them.

The first two chapters of the book present a scientific framework for written study plans, with a brief overview of the philosophy, sociology, and history of science. These chapters attempt to draw practical lessons from these disciplines for those actually planning and doing research. The hypothetico-deductive approach to science is emphasized, as is the utility of an adaptive management and research approach to many environmental and natural resource problems.

The next five chapters discuss in detail the components of written study plans and methods for reviewing and presenting them. These chapters together cover the contents to be expected in any competent written study plan in the order they are listed below.

The final chapter discusses the impact of planned research on professional practice and public policy. Enlarging the general fund of human knowledge is not the only contribution environmental and natural resource research is expected to make. The final test of the efficacy of research planning is the positive effect the results have on activities in the field, on profit and loss, and on the creation of effective laws and administrative rules.

STUDY PLAN CONTENTS

A good study plan will include all the elements listed below, and will be concise. A good study plan will also follow the format that will be most helpful in accomplishing its purpose. It may be in the form prescribed by a granting agency or an employer. The format most acceptable in the disciplinary culture may be used. But a good plan will contain all the elements listed, regardless of their arrangement or specific title.

A title concisely conveys the content and intent of the plan.

1. The author's name, affiliation, and contact information, along with the date and number of the draft (if it is such).

2. An abstract that is not more than 3 percent of the length of the whole plan.

3. A summary statement of the research problem that says clearly why the research is being done, and who will care about the result and why.

4. A statement of the objective of the proposed research, with criteria by which completion and success can be judged.

5. Hypotheses that are testable.

6. A list of variables and sources of variation that bear on the objective and hypotheses.

7. The study design and intended methods of analysis.

8. Field, laboratory, and computational procedures to be employed and why.

9. A list of intended products, including reports, publications, and constructions.

10. A budget for the life of the project, presented so that sources and uses of funding are clear.

11. A schedule of the activities needed to complete the project.

12. Literature cited in the body of the plan.

13. Appendixes such as detailed methods, literature reviews, maps, and supporting information generally.

I discuss each of these written plan elements in a later chapter. The important point is that a complete plan contains all of them, with no exceptions. Even if the chosen plan format doesn't specify them all, they should be included at least in the author's version, because the primary audience for any plan is the author herself. Including all of the elements causes the author to think about them together, carefully, before embarking on other research activities. Without this reality check, it is much easier to be overly optimistic about anything from hypotheses to budgets. Most of all, answering the "why and who cares" questions reassures the author that this is not a project "not worth doing at all." Here is a complete study plan outline:

A Study Plan Outline.

I. Title

II. Author and Date of Draft, Author Contact Information

III. Abstract (3 percent of total plan length)

IV. Problem Statement

  A. Summary of relevant literature (a lengthy literature review and synthesis can be presented in an appendix)

  B. Context of the research (geographic, institutional)

  C. Problem definition, five elements or "gap in knowledge"

  D. Projected outcomes and findings

  E. Relationship to other studies; cooperation

  F. Summary of costs and benefits

V. Statement of Question, Objectives, and Hypotheses

    A. Overarching research question, derived from problem statement

    B. Intended outcome of this piece of research (usually to test hypotheses and define relationships), including an estimate of time and resources needed

    C. Specific mechanistic hypotheses, tests in summary, populations to which the hypotheses are to be applied

VI. Methods

    A. List of variables and sources of variation, sorted by independent and dependent variables, with reasons for their selection

    B. Other sources of variation and how they will be dealt with

    C. Study design and analysis, including models, statistical tests if used, detailed analytical procedures, graphs of potential outcomes

    D. Field, laboratory, and computational procedures, in sufficient detail so that someone other than the author could do the study

VII. Budget and Schedule

    A. A comprehensive three-column budget for the duration of the study

    B. A schedule of tasks with initiation and completion target dates, with designated responsibilities and reporting requirements and dates

VIII. Reports and Publications

    A. Intended disposition of research results, in terms of audience, publication type, and timing

    B. Fiscal, accounting, and procedural reporting requirements and how they will be met

IX. Literature Cited and Appendixes

TWO

Scientific Method

Scientific research can be classified as basic or applied, natural or social science, or in other ways, but several axioms or assumptions are at the base of all science. The most universal of these are observation, experiment, measurement, logic, order, honesty, proof, and repeatability. These classifications are most frequently applied in the testing of mechanistic (specifying cause, pathway, and effect) hypotheses in ways that exclude certain interpretations of the results of the tests. Understanding the steps in the cycle of research activities (the scientific method), as well as understanding the axioms, is the first step in preparing good research plans.

KINDS OF SCIENTIFIC RESEARCH

Scientific research is a learned skill. Some people appear to have a greater aptitude and affinity for it than others, but it is clearly a human activity, since animals don't do it. As such, it is affected by all human attributes, positive and negative, and is not, in that sense, very different from other structured human activity. Only in the collective sense of many together achieving a level of knowledge that individuals cannot alone is science special. Often a difference is drawn between basic and applied research. In the sense of the end toward which the research is directed it is a useful distinction. Basic researchers strive to answer questions about the nature of the world and universe. These questions usually come from science itself, having been suggested by prior research or theory. Applied researchers try to solve problems that usually arise, at least in their original form, outside science. Whether trying to develop a commercial product or devise a way to save an endangered species, the success of applied research is judged by whether a human problem is solved as a result.

Another popular science classification is into natural and social science. This distinction can turn on the object of study, with social scientists studying people and natural scientists studying other things. But that doesn't really work, given the huge effort by natural scientists to study people, driven largely by medical concerns. Social science is sometimes pejoratively called "soft" science, on the grounds that it does not include the same kinds of experimental protocols and hypothesis testing that natural "hard" science does. It has also been frequently charged that social sciences are at least some of the time rooted in political ideologies and therefore less objective than natural science. It is probably true that all science is influenced by the society around it, including political movements and perspectives. In the most basic view of science methods (see below), however, there is no need for social science to be regarded as less rigorous or credible.

Another difference within science that is often discussed is between quantitative and qualitative research. Qualitative research depends on the comparison of attributes of researched objects that cannot be represented by numbers. Despite the dictum of Lord Kelvin to the effect that if you can't quantify something you don't know anything about it, there is now a large body of methods applicable in science labeled as qualitative. Indeed, most quantitative studies depend in some way on descriptions of qualities like color, shape, texture, or narratives of behavior.

These distinctions are important for some purposes, but they are not very important from a planning and execution point of view. The basic rules of science, and therefore the basic planning procedures, validation criteria, and presentation modes, are strikingly similar across these boundaries, even though subjects and methods differ greatly.

Some argue that science is no more objective or "special" than any other form of belief system, and therefore of no more validity or utility. Scientific research is indeed subject to all possible human failings and inadequacies. But science has done more to render human failings apparent, and to hold them to account, than any other approach (such as in folklore or religion) to the creation of knowledge. To take one sector, science has done much to improve how we manage the human environment, and it is hard to see what the substitute would have been. "Common sense" has often produced disastrous solutions to environmental problems. For example, the common-sense prescription for waste disposal was "out of sight, out of mind, no harm done." In any event, scientific research is the surest and most efficient way yet devised to increase human knowledge. From a research-planning point of view there are only two kinds of research: good and bad. Good research achieves its objective, often, though not always, by testing a mechanistic hypothesis (see below). Bad research misses its objective and allows no valid conclusions to be drawn, or only trivial ones. Very often, though again not always, research that is labeled observational or monitoring, or involves collection of baseline data falls into the "bad" category. This is not, however, because it is intrinsically bad to observe, monitor, or collect data. Rather, it is usually because the activities are poorly planned, and the basic methods of science thus are not given proper attention. If insufficient thought is given at the outset to the ultimate use of the intended results, one often finds that there isn't one.

ELEMENTS OF THE SCIENTIFIC METHOD

An overview of the general method of science often includes the elements listed below. Many have commented that there are many approaches to science that differ by discipline and subject, and thus there are many "scientific methods" (for example, see Black 1977). Also, the listed elements are common to many other human activities. The uniqueness of the scientific method lies in their rigorous and creative application together. If all of the elements aren't present, bad science is likely to result. Written plans allow the researcher and reviewers to check to see if all of these bases have been considered and covered at the outset.

The basic elements of science are:

Observation: the unbiased perception and recording of objects and events in a form communicable to others;

Experiment: creating or finding structured situations and events that are especially suited to unbiased observation and the drawing of conclusions, usually through the test of a hypothesis;

Measurement: the accurate, usually quantitative description of objects and events with reference to widely agreed-upon and verifiable standards;

Logic: the use of inductive and especially deductive thought processes to reach conclusions in ways that make the individual steps in the process clear and verifiable;

Order: the assumption of cause and effect, so that objects and events have a single, knowable history and a predictable future within the limits of current knowledge;

Honesty: the covenant that observations, experiments, measurements, and logic found to be flawed or incorrect will be discounted and discarded, as will disproved hypotheses;

Proof: the continual submission of observations and predictions to further tests of the kind that can prove them flawed or false, if indeed they are; and

Repeatability: the belief that repeated causes and conditions will have at least approximately repeated consequences.

Some important terms used to describe the scientific method are defined below.

Bias is the skewing of observation and measurement so that true or intrinsic properties are not seen or measured correctly. It can arise in sampling through non-randomness and in observation through willful distortion or faulty technique, for example.

Logic is the application of disciplined, rule-bound, and repeatable methods of reasoning to structured problems so that defendable and explainable conclusions can be drawn. Common evasions of logical thinking and analysis include: (1) appeal to authority, as in "what I say is true because Professor X says it is true"; (2) ad hominem, "to the man," as in "what Professor X says is wrong because he is a terrible person"; (3) appeal to emotion, as in "what I say is true because I feel so strongly that it is"; (4) appeal to ignorance, as in "what I say is true because you can't disprove it," with the statement offered without evidence of how "it" could be disproved; (5) begging the question, or the offering of a circular argument, as in "what I say is true because I have assumed preconditions that make it true" (read about the stock market with this one in mind); (6) diverting the issue, as in "what I say is true because elk cross the prairie," when the asserted proposition has nothing to do with elk or prairie; and (7) special pleading, as when only evidence supporting a conclusion is presented, when evidence to the contrary also is readily available (Ruby 1950).

Induction is generalization from specific cases or evidence. A professor is riding a train and his companion looks out the window. He sees first one, then two, and finally three sheep that have been sheared. He remarks to the professor, "I think we can conclude that all the sheep in this county have been sheared." The professor looks up from his book and says, "at least on one side."

Deduction is the extraction of specifics from general cases or evidence. The syllogism is a simple form of deductive logic, and it has three elements: a major premise, a minor premise, and a conclusion. To work, the premises must be correct, and correctly related to each other, for a valid conclusion to be reached. Spot the flaw in each of these fanciful syllogisms about "My Dog Spot" (fanciful, but nevertheless similar to more complex instances of flawed reasoning found both inside and outside science):

> All dogs have tails
> Spot has a tail
> Therefore, Spot is a dog

> Only dogs have tails
> Spot has a tail
> Therefore, Spot is a dog

> Some dogs have tails
> Spot has a tail
> Therefore, Spot is some dog

Mechanistic hypotheses are predictions that explain not only what is expected to happen under specified conditions, but also why and how it will happen. Thus, hypotheses stated in this way describe the mechanics of its potential realization, as in "the rabbit will appear from the hat because it was concealed in a hidden compartment in the hat's crown and then propelled upward by a spring mechanism." The simple statement of a null hypothesis—that is, that there will be no difference in experimental objects under a given set of conditions—is a logical necessity but does not constitute a mechanistic hypothesis because it offers no explanation.

Fraud (in science) is most obviously and frequently the willful altering or faking of research data or results to suit the purposes of the researcher. Science at its core is a form of rigorous honesty, and thus fraud strikes at the heart of good science.

A number of axioms or "durable assumptions" that defy proof except by continual effective use also bound science, but less rigidly than those above. Words like *belief* and *covenant* can be used to describe those elements of the scientific method that are assertions and as such not subject to either theoretical proof or complete validation by experience. Their justification is only that they seem to work. At least in theory, there could be a future time when they might not.

The most important of these are:

- Authority and revelation do not establish truth, and truth is only that which is not currently proved false, assuming there has been a rigorous attempt to do so

- Observation and description are necessary to the use of the scientific method but are not sufficient to place an activity inside the scientific method, and that observation should use unbiased, quantitative, and instrumented (when possible) methods.

- Prediction and control as proof of understanding, through knowledge of cause and effect, are the primary objectives of the use of the scientific method.

- Abstraction, analysis, and synthesis are necessary to the scientific method, where modeling (see Chapter 4) is the abstraction of essential elements needed for a given analysis, analysis is the disassembly of a system into understandable components, and synthesis is the assembly of known components into an understandable whole.

- Mechanistic hypotheses are the soul of the scientific method, and their construction and testing are the central creative act in the application of the scientific method.

- Science is a collective human activity subject to all human frailties, quirks, cultural biases, and general cussedness, but, because it is collective and internally critical, tends to be self-correcting.

There probably never will be complete agreement on the axioms underlying the scientific method as it is used in practice, given the sprawling nature of science. Those above have proven useful in understanding and practicing scientific research and provide a checklist to use while writing research plans. However, the method finds its real validation in its results. One science observer has said that civil engineering research was probably the most easily and publicly validated kind of research, since if a bridge falls down either the science or its application was flawed. Perhaps the most important thing about the assertions or assumptions at the base of science method is that even if they are all perfectly observed in practice, the method still tends to work.

Many studies indicate that current science uses knowledge from a huge range of disciplines and times (for example, see Illinois Institute of Technology 1968). Thus to isolate any piece of research, or to ascribe to it a precise beginning and end, is technically incorrect. One doesn't have to be "completely original" to do science; indeed it is impossible to be globally original and to stay within the boundaries of science. Nevertheless, it is important, particularly for beginning researchers, to get a sense of order, and what comes first, in planning a specific research episode. A simple set of steps, which actually describes a cycle, can illustrate the structure of any piece of research in the planning stage:

1. Identification of the preliminary data and observations that suggest a question to be answered. These often come from direct observation by the researcher of an unusual or unexplained occurrence, or from a conflict or gap in the scientific literature, or from an urgent question raised outside science, or a combination of these.

2. Formulation of the problem. What is the specific question to be

answered, why is it significant, and what are the possible outcomes of answering the question or solving the problem? (See Chapter 4.)

3. Further gathering of information by observation or reading that allows definition of the components and boundaries of the question. Intensive review of the literature is a usual part of this step.

4. Formulation and selection of mechanistic hypotheses that, when tested, will answer the question (see Chapter 5).

5. Deduction of the implications of the hypotheses that will allow direct tests of them to be constructed.

6. Structuring and doing the hypothesis tests, usually in a designed experiment format (see Chapter 6).

7. Drawing and communicating conclusions, including the formulation of new hypotheses and tests.

Given this sequence, it is clear that science, as practiced, is cyclic and open-ended in the sense that no end point is reached, even though new information is gained. This cycle is discussed in greater detail by E. David Ford (2000).

Some have speculated that most great scientific discoveries possible have been made already (see, for example, Horgan 1997), at least in physical science and biology, and that the basic structure and content of the universe is understood about as well as it ever will be. This in itself is an assertion about the future that can not be scientifically evaluated, but even if true, it probably also predicts a golden era of the application of the methods and talents of science. With such a firm and unchanging framework, and a workforce largely released from cosmic pursuits, it should be possible to use science to solve a whole host of sub-cosmic but vexing human problems, especially environmental ones.

Indeed, many of the most important human problems lie in the realm of the environment and the management of natural resources. Better information and predictions about climate, oceans, rivers, soils, forests, and human interactions with them would be immediately interesting and useful, even if they do not lead to fundamental new insights about the nature of the universe. True, this will be, in the eyes of some, lesser science. But whether it is done, and how well it is done, may determine whether humanity has a positive future, or any future at all.

THREE

Useful Views of Science

Conducting scientific research is a learned skill. Certain principles apply across all kinds of research, and their observance, or lack thereof, determines research quality. Philosophers and historians of science have argued about these principles for centuries, but a number of their ideas apply particularly well to written plans. Kenneth Boulding, Karl Popper, and Thomas Kuhn had diverse views of science, but a practical and useful whole can be extracted from their writings. Science is an intrinsic human activity, subject to all human failings, but, because of its intrinsic nature, it is accessible to all humans. Normal science often operates within generalizations called paradigms. Science proceeds unevenly, with periods of normal science between revolutions or paradigm changes. Promulgating hypotheses and trying to disprove them are a mainspring of normal scientific advance. The more ways there are to disprove a plausible hypothesis the better it is. Nontrivial hypotheses that survive rigorous attempts at disproof over time are eventually called theories and laws. Knowing these views of the science process will help you to communicate research plans to broader audiences.

The literature of the history, sociology, and philosophy of science is enormous and often impenetrable to nonspecialists in those disciplines; however, it contains valuable lessons for the research planner. At the least it is instructive to try to fit your planned piece of research into the various views of how science happens.

BOULDING

Kenneth Boulding was an academic economist without a Ph.D. This, perhaps, accounts for his often original, non-mainstream views. His view of the origin of science rings true (Boulding 1980). Like others who have described science in relation to other human activities, Boulding looked for similarities and differences between science and other modes of thought and action. He saw science as a product of noogenetic (transmitted by the replication of knowledge structures—that is, books and teaching) evolution, although it is structured in biogenetically produced human brains. Thus, he saw science as a cultural product, the result of a general human tendency toward scientific method because "the principal business of the human mind . . . is fantasy. Nevertheless, part of the structure of (fantasized) images is their labeling with our belief in their degree of reality. From these images we form expectations. When an expectation is fulfilled, it reinforces the image" (832). He thus suggested a natural origin of the process of hypothesis formation, since hypotheses are, in their simplest form, expectations that can be tested. Boulding concluded, "knowledge increases not by the matching of [mind] images with the real world (which Hume pointed out is impossible), that is, not by the direct perception of truth, but by a relentless bias toward the perception of error. This is as true of folk knowledge as it is of science" (836). Error detection has been described as central to science by Popper (see below) and many others. Things can be proved wrong in science, but they cannot be proved right except in the sense that they haven't (yet) been proved wrong.

Just as Boulding saw a natural origin for scientific thought, he also saw a continuum of ways of thinking, rather than a sharp boundary between science and non-science. He asserted that science was a "fuzzy set" within the vast domain of human thought. This is a useful view to keep in mind when contemplating some of the more didactic views of science. Look at the dialogue below and reflect on what its content and tone say about science as well as those whom scientists see as the enemy.

A conversation with Dr. Noitall, "Two Plus Two Equals Five"

SCIENCE. Dr. Noitall, you are the world's greatest authority on public relations, the man who could get Brezhnev elected in Orange County, the man who could sell crepes suzette as the breakfast of champions.

DR. NOITALL. A vast understatement of my true worth.

SCIENCE. We have come to ask you why scientists seem to have such a poor image.

DR. NOITALL. How can you possibly ask such a simple-minded question? You are the people who have brought us nuclear war, global warming, and acid rain. You enjoy dissecting frogs, and you keep mice and rats in little cages instead of killing them with warfarin, as any decent softhearted farmer does.

SCIENCE. But we're not in favor of nuclear war.

DR. NOITALL. You discovered the atom. You had your chance to stick with phlogiston and you didn't do it.

SCIENCE. We are not in favor of global warming. In fact, we're the ones who alerted the public to this danger.

DR. NOITALL. That shows the naiveté of scientists. The Bible records the execution of messengers who brought bad news. We no longer execute such people, but we certainly don't have to like them. You tell me I have to give up my air-conditioned gas guzzler for an itsy-bitsy, crowded, four-cylinder, nondescript vehicle, and expect me to like you.

SCIENCE. But it is our job to tell people when $2 + 2 = 4$.

DR. NOITALL. That's exactly where your views are wrong. A recent poll shows that 50 percent of the people think $2 + 2 = 5$, and almost every network agrees with them. Those people have rights, they believe sincerely that $2 + 2 = 5$, and you take no account of their wishes and desires. Simply imposing $2 + 2 = 4$ on them is not democracy.

SCIENCE. But there is really no serious scientific disagreement on the question.

DR. NOITALL. That is exactly where the problem comes in. The Establishment is lined up monolithically on the side of $2 + 2 = 4$. A dissident scientist representing the $2 + 2 = 5$ school cannot get this article published in Establishment Journals. Peer review is utterly unwilling to look with favor on such viewpoints. Granting agencies do not encourage people who believe $2 + 2 = 5$ to serve on their boards.

SCIENCE. We can't take seriously people who make emotional rather than scientific arguments.

DR. NOITALL. That reflects a condescending attitude toward those who did not have the privilege of having an advanced education. Prominent political groups have already supported enactment of legislation, even if it is scientifically inaccurate, as long as the public wants it.

SCIENCE. Then how can we go about changing public opinion?

DR. NOITALL. In the first place, you could stop having funny-looking people in glasses and lab coats appearing on television and before legislative committees. Pick real successes, such as actors, actresses, and rock singers, and let them present your side of the story. Making large amounts of money and being handsome is evidence of success in modem society. Such people are much more likely to understand big subjects than professional types. And stop defending unpopular causes. It is now understood that atoms, asbestos, CO_2, and cholesterol are bad. Attempts to say that they're all right in small doses is only going to get you an image as an apologist for bad guys.

SCIENCE. But even if we pick attractive spokespersons, how can we convince them of the truth?

DR. NOITALL. That is another misconception that you are going to have to get rid of. There are two truths in this world: one of the laboratory, and the other of the media. What people perceive as the truth is truer in a democracy than some grubby little experiment in a laboratory notebook. A stubborn insistence on the facts instead of the people's perception of the facts makes you look heartless and disdainful. You are going to have to come out as more reasonable and accommodating, as more benevolent, kindly, and pliable, willing to give and take, empathizing with the public's tears and frustrations.

SCIENCE. So how do we handle the $2 + 2 = 4$ problem?

DR. NOITALL. I'd suggest you start by conceding that $2 + 2 = 4 \ 1/2$. (Koshland 1990: 1381)

POPPER

Two of the twentieth century's most durable and controversial commentators on the nature of science are Karl Popper and Thomas Kuhn. Their writings have been criticized and elaborated by others since publication, but their original writings contain much that is useful to the research planner. Both developed densely argued insights intended to illuminate the "real" nature of science. Both were skeptical of the descriptions of science found in textbooks. It is important to understand the basic thrust of their arguments for two reasons: (1) they offer practical help to the research planner, and (2) their thinking, though controversial, has shaped discussions of what science is and how it works and continues to do so.

In *Conjectures and Refutations* (1962), Popper stated his central questions as "When should a theory be ranked as scientific?" or "Is there a criterion for the scientific character or status of a theory?" He went on to explain, "The problem that troubled me at the time was neither, 'when is a theory true?' nor 'when is a theory acceptable?' . . . I wished to distinguish between science and pseudo-science; knowing very well that science often errs, and pseudo-science may happen to stumble on the truth" (33). This quest was and remains exciting to a wide array of people interested in science. If a surefire method exists to tell science from pseudo-science it would be of obvious utility to the planners, reviewers, and users of scientific information.

His book title, *Conjectures and Refutations,* is an adequate, if condensed, statement of his view of the method of discrimination he proposed. After analyzing cases he chose to exemplify both science and pseudo-science, he concluded:

> It is easy to obtain confirmations or verifications, for nearly every theory—if we look [only] for confirmations. . . . Confirmations should count only if they are the result of risky predictions; that is to say, if unenlightened by the theory in question, we should have expected an event which was incompatible with the theory—an event which would have refuted the theory.
>
> Every "good" scientific theory is a prohibition: it forbids certain things to happen. The more a theory forbids, the better it is.
>
> A theory that is not refutable by any conceivable event is nonscientific. Irrefutability is not a virtue of a theory (as people often think) but is a vice.
>
> Every genuine test of a theory is an attempt to falsify it, or refute it. Testability is falsifiability; but there are degrees of testability: some theories are more testable, more exposed to refutation, than others; they take, as it were, greater risks.
>
> Confirming evidence should not count except when it is the result of a genuine test of the theory. This means that it can be presented as a serious but unsuccessful attempt to falsify the theory.
>
> Some genuinely testable theories, when found to be false, are still upheld by their admirers—for example, by introducing ad hoc some auxiliary assumption, or by reinterpreting the theory ad hoc in such a way that it escapes refutation. Such a procedure is always possible, but it rescues the theory from refutation only at the price of destroying, or at least lowering, its scientific status.
>
> One can sum all this up by saying that the criterion of the scientific status of a theory is its falsifiability, refutability, or testability (ibid., 33).

Popper thus suggested a potentially quantitative scale for rating the scientific quality of theories; the more testable they are, the more scientific they are. Greater risk of falsifiability brings greater scientific credibility if the risk is taken (the test is made) and is survived. In his seventh point in which he described the "conventionalist twist," the ex post facto change that allows a theory to survive, he paralleled Kuhn. Kuhn (1970) observed that anomalies, results that test and seem to refute a theory, are ignored or explained away in "normal science" until a "scientific revolution" occurs (see below). Popper also paralleled Boulding's view of science as part of the normal working of the human mind as it checks fantasies against perceived reality. Popper said, "I realized that . . . myths may be developed, and become testable; that historically speaking all—or very nearly all—scientific theories originate from myths, and that a myth may contain important anticipations of scientific theories. . . . I thus felt that if a theory is found to be non-scientific, or 'metaphysical,' . . . it is not thereby found to be unimportant, or insignificant or 'meaningless' or 'nonsensical.' But it cannot claim to be backed by empirical evidence in the scientific sense—although it may easily be, in some generic sense, the result of observation" (Popper 1962: 38).

Popper said that science happens because scientists, who may take their ideas and inspiration from outside science, make risky predictions and try to refute them. This implies that the best hypotheses are those most exposed to testing. Theories are just hypotheses that have been subjected to rigorous test and not, at least yet, been refuted. It also implies that without prediction and test, whatever credibility accrues to an idea because it is "scientifically valid" or just "scientific" is denied.

This implication is especially important in the vast areas of environmental science, ecology and biology, and social science in which observation of populations and systems not subject to structured, "human-made" experiments (see "Finding Experiments" in Chapter 6) is the source of test and refutation. Unless the observer can convince a skeptical third party that the elements of "risky prediction" and serious attempts at refutation are present, they will not find their results viewed as more than interesting myth by those who accept Popper's criteria.

KUHN

Kuhn, in an inspired doctoral dissertation that later became a book, *The Structure of Scientific Revolutions* (1970), said that the conventional view of science as a body of knowledge that builds on itself is not just an over-simplification but is wrong. Past views of science, he said, had been constructed after the fact, largely in books with a "persuasive and pedagogic" aim. He saw science working by the rise and fall of overarching syntheses, which he called "paradigms." The currently accepted synthesis, or paradigm, sets the boundaries of scientific inquiry, indicates scientific questions to be answered ("puzzles to be solved"), and specifies appropriate methods to use. "Normal science" occurs within a paradigm and solves the puzzle set by the paradigm, and its advances are limited by the boundaries of the paradigm. A scientific revolution occurs when a new paradigm replaces an old one because the old one cannot bear the weight of accumulated anomalies and unsatisfactorily explained results. The boundaries then change, normal science proceeds to solve different puzzles, and the foundation for a new revolution (the eventual fall of the new paradigm) is laid. Perhaps his most controversial view is that the new paradigm is not necessarily intrinsically better (more beautiful, more true) than the old one. It has replaced the old one simply because it is a more complete and consistent synthesis of what is known and accepted.

Kuhn thus viewed science as an open-ended, poorly defined, but highly efficient endeavor. Its efficiency, in his view, lies in the construction, use, and replacement of paradigms. Specialization within a discipline is possible because each scientist doesn't need to understand the whole field. The paradigm takes care of that, as well as the definition of holes in knowledge and the setting of rules by which they can be filled. Paradigms also make communication more efficient, by sharply restricting the breadth each individual attempt at communication must have. Once a paradigm is established, Kuhn (1970: 20) said, "No longer will . . . researches usually be embodied in books . . . addressed, like Franklin's *Experiments* . . . or Darwin's *Origin of Species* to anyone who might be interested in the subject matter of the field. Instead they will usually appear as brief articles addressed only to professional colleagues, the men whose knowledge of a shared paradigm can be assumed and who prove to be the only ones able to read the papers addressed to them." Thus, the paradigm deftly defines both the audience and the language used to reach it.

Kuhn saw important differences among fields of science in their development of paradigms and suggested that their development and adoption were good indicators of the presence of true science. "Except with the advantage of hindsight, it is hard to find another criterion that so clearly proclaims a field of science." He gave a clear account of what happens without paradigms. "In the absence of a paradigm or some candidate for paradigm, all of the facts that could possibly pertain to the development of a given science are likely to seem equally relevant. As a result, early fact gathering is a far more nearly random activity than the one that subsequent scientific development makes familiar. Furthermore, early fact gathering is usually restricted to the wealth of data that lie readily to hand. The resulting pool of facts contains those accessible to casual observation and experiment together with some of the more esoteric data retrievable from established crafts like medicine, calendar making, and metallurgy. Because the crafts are a readily accessible source of facts that could not have been casually discovered, technology has often played a vital role in the emergence of new sciences" (Kuhn 1970: 15).

Thus Kuhn, like Popper, was concerned with the boundary between science and non-science, and he suggested a test—that is, the degree to which a common paradigm has emerged. When he wrote, he was unsure about the status of parts of biology with respect to this test, and he said, "it remains an open question what parts of social science have yet acquired such paradigms at all" (ibid., 16).

USING VIEWS OF SCIENCE IN WRITING PLANS

The writings of Kuhn and Popper have been thoroughly reviewed and criticized, from both philosophical and scientific points of view. But both philosophers have raised seemingly permanent questions about the nature of science. Practitioners of science, particularly those attempting to answer environmental and natural resource questions, can benefit intellectually and in daily activities from reflecting on the questions: "what constitutes science and how does it differ from other activities?"; "what paradigms, if any, am I operating within?"; "how risky are my predictions (hypotheses), and what would constitute true refutation of them?"; and, particularly, "how much of my information is derived from what Kuhn called crafts (professional practice) and what does this mean about its scientific credibility?" Credibility within science is established by the norms of science, not practice. Sadi Carnot formulated an early version of the second law of thermodynamics but was an engineer trying to design better steam engines. Thus his work was slow to be incorporated into the mainstream of physics. Practice and science ideally nurture each other, like the shoot and root of a plant, but methods and proof move uneasily between them.

Attempting to answer the questions above explicitly in the planning stage will improve the effectiveness with which any individual piece of research is carried out, and it will enhance the credibility of its eventual result.

One interesting exercise is to attempt to identify paradigms that guide research in environmental subjects, such as wildlife biology, forestry, and ecology. Because wildlife biology and forestry are not only research areas but long-standing crafts, one might speculate that they are, in Kuhn's terms, in a pre-paradigm state. They rely heavily on information developed from the practice of their related profession and on paradigms borrowed from the full range of more mature sciences.

Ecology, on the other hand, seems to be a special case (Schrader-Frechette and McCoy 1993). Although it draws heavily on the physical sciences and other parts of biology, ecology is different enough from other parts of biology to be designated separately, with scientists referring to "biology and ecology." It often is also segregated into a separate department academically. It also has no closely allied craft of long standing (conservation biology and restoration ecology are new, and they still have the flavor of science rather than craft), although information from ecology is used in an array of human activities from pollution abatement to farming and forestry. Ecology is a relatively new science, but its "differentness" probably lies most clearly in its insistence on including a holistic view of nature, and its ingrained suspicion of reductionist science, even while using it. In any event, ecology, "the study of the relation of organisms to their environment," is central to even the most applied research on environmental and natural resource research topics. Three terms are used to define the relationship between ecology and the rest of science:

- *Holistic* is used to describe a view of nature and science that deals in whole entities, possessed of all their attributes, and implies the embrace of complexity and multiple interactions among components. It is associated with whole-organism and whole-system approaches to understanding nature and humans' place in it.

- *Reductionist,* often used pejoratively by holistic people, describes the view that the simplest possible systems that can contribute answers should be the objects of study, and that pictures of whole organisms and systems can be built from studying their components. This label is often applied to laboratory scientists and other students of "components."

- *Realism* is the recognition that to understand nature and solve problems it is necessary to look carefully at both the whole and its parts.

FOUR

Stating Problems and Objectives Clearly

The most important steps in planning research are setting research priorities, defining the problem research will address, and writing clear objectives. Problems can be defined as gaps in knowledge or in terms of choices decision-makers must make. A gap in knowledge is most often identified from the literature or a new observation. The decision-maker method is most often used in applied research that aims to help people outside science. The latter method has five important components that, when fully described, constitute a complete problem statement. Those are: the decision-maker, his or her objective, alternative paths to achieve the objective, a statement of doubt about which path to choose, and the context in which the decision-maker operates. Environmental problems are often complex and are seldom completely solved by one study or by answers obtained from modeled solutions. Once the problem is defined, objectives, including any hypotheses to be tested, are written, and they should clearly indicate the intended outcome of the research. A good objective also indicates the time and money it will take to complete the study and gives a minimal statement of method. Titles are condensed objectives and are important to the thought processes of the researcher and to reviewers, so they should be crafted with care.

SETTING RESEARCH PRIORITIES

At the highest level, research priorities are set by the political interplay between science and society, and they are expressed most often through governmental or market allocation of funds. In that sense, the individual scientist planning a single study has little influence on overall priorities and doesn't have to worry about them except to know that funds are or might be available for the planned work. Within the macro-environment of requests for proposals, government and corporate objectives, researcher peer pressure, and academic fashion, however, there is considerable latitude for individual choices. One way to approach the problem of choice is purely scientific. If a paradigm exists to be consulted, what does it point to as the most important holes to fill or puzzles to solve? Or, put another way, what does science say is most important to do? The sources of information about what "science" wants can come from synthesis or review articles, the researcher's own evaluation of the literature, discussions with colleagues, and direct observation. Most often the triggering situation is the need to resolve a conflict or contradiction, new information that suggests that an existing theory or hypothesis needs further testing, or a hypothesis that needs further scrutiny.

Another approach is to do a form of cost-benefit analysis. This involves, on the benefit side, determining:

- How many people have the problem?

- What is or will be the cost to them of wrong decisions?

- How much would the new information do to prevent wrong decisions?

- What comparative advantage does the researcher have that makes problem solution more likely than if another person or research organization pursued it?

On the cost side, there are three major things to determine:

- How much will the research cost in time and dollars?

- What is its opportunity cost (what doesn't get done if this is done)?

- Can the research be financed at all (what is the probability of a good plan being funded)?

The most important thing in a written plan is to describe why you have chosen to do this research, and to give a sense of why this research is a high priority.

THE GAP-IN-KNOWLEDGE MODEL

The simplest model to describe the specific reasons you have chosen to work on your problem is the "gap-in-knowledge" approach. Something is not known that needs to be known and it can be expected to be known if the right research is done. Paradigms (discussed in Chapter 3), the existing syntheses of knowledge in the field, tell what is known and what needs to be known (gaps in knowledge) to gain a fuller understanding of the part of the universe the field addresses.

Sources that allow such gaps to be identified in a practical sense include textbooks, annual reviews of the field or subfield, and scientific articles that propose new directions. The writings that bound the gap need to be specified, and their relation to one another needs to be explained. This approach is usually, but not always, confined by the scientific literature. Sometimes the researcher makes a personal observation and clarifies such a compelling need to know more that a major effort to pursue the knowledge is warranted. The discovery of biochemical pathways of photosynthesis other than those of the Calvin cycle is a case in point. For a long time, most research on the biochemical steps in photosynthesis had been done on only a few plants. The Calvin pathway was thought to be the only one, but then it was observed that certain plants (some commercially important grasses, such as sugarcane) gave results that would not be expected if the Calvin cycle was the pathway by which carbon moved from the atmosphere into plant structure and energy transformations. When these anomalies were pursued directly, a whole new set of photosynthetic and respiratory events that had great theoretical and practical implications was discovered.

In the absence of a guiding paradigm or a startling and significant new observation, the gap-in-knowledge mode of problem definition tends to lose value as a credible way to convince others that your problem is significant and high priority. There are infinite gaps in knowledge, and thus the fact that there is a gap is of little significance by itself.

THE DECISION-MAKER MODEL

A method of problem identification, description, and prioritization that lends itself especially well to environmental and natural resource research depends on identifying a decision or class of decisions that is difficult or impossible to make based on current information. This decision-maker model requires that five elements of the problem be explicitly and carefully described before it can be said to be one that research can help resolve (Ackoff 1962, Stoltenberg et al. 1970).

This method depends on close communication between the researcher and those making decisions to be aided by scientific research. Thus, it is almost always associated with applied research. Because most environmental and natural resource research is done with a practical outcome in mind—whether it is, say, increasing water yield and quality or saving an endangered species—it fits environmental science well. It can be applied whenever a problem of human choice exists, however, including when choices are made inside science for the purpose of seeking "pure" knowledge, with no application in mind.

The five problem elements that must be specified to have a complete problem definition in this sense are:

1. *The decision-maker.* This is the person or class of persons that has the problem of choice. Usually, in fact, it is a group of people; for example, National Park managers who have to deal with large elk populations, or forest managers who want to improve the quality of the hardwood trees they grow. A decision-maker is rarely a single individual, but it is important that the researcher be able to name one or more of the people in the group said to have the problem and to be able to verify that they agree that they have it.

2. *The decision-maker's objective.* This is what the group with the problem wishes to achieve. Perhaps the Park managers want to ensure that elk populations are not threatening other resources in their parks by browsing certain plant species too heavily. The forest managers may want to increase the proportion and growth rate of a certain species, red oak, for example, in their forests. The important thing is that it is the decision-makers' objective, not the researchers', at this point.

3. *Alternative ways to achieve the objective.* These are possible pathways to the end the decision-makers have in mind. The Park managers, for example, could reduce elk numbers, protect the threatened vegetation in other ways, or both. The forest managers could remove other species to favor the oak with more growing space and to make it a greater proportion of the trees in the forest, fertilize the oak to increase its growth rate, or plant more oak in forest openings. The key is that there must be different ways to pursue the objective or there is no problem of choice, although it may be that no cur-

rently available option is looked upon with favor. In this case, research can pursue new options or new combinations of old ones, but each current option must be carefully and fully described to have completed this problem definition element.

4. *Doubt about which alternative to choose.* If it is clear which alternative is most desirable, there is no doubt and, in this sense, no researchable problem. Though this seems obvious, much applied research is almost characteristically shaped by old, solved problems. This in turn means that much effort is directed to refining answers beyond the need of decision-makers rather than identifying and solving new problems.

5. *Context:* No decision-maker or problem exists in a vacuum. Their surroundings affect both problem and solution. Does the Park manager with elk devouring all the vegetation have neighbors or neighbor influences that affect the elk or the vegetation they browse? Are there economic or regulatory constraints on the forest managers that might keep them from thinning other species to favor the oak they may desire? All influences in the milieu of the decision-maker that might affect the choice among alternatives need to be specified and considered in this element of the problem definition.

The decision-maker model of problem definition is simply a more focused case of the gap-in-knowledge approach. Its virtue is that it states clearly who has the problem (the decision-maker) and why (the problem of choice) and thus answers the "why this particular research now?" question more rigorously. This approach is seldom explicitly taken in "pure" research, but it could be. In that mode, the decision-makers are other scientists seeking to advance knowledge in their field. Their generic problem of choice is which hypotheses to retain and use as they go forward. This is not so different from using the "strong inference" method described by Platt (1964) (see Chapter 5).

"SQUISHY" PROBLEMS

Strauch (1975) wrote a paper called "'Squishy' Problems and Quantitative Methods" that, despite its relative antiquity and reference to Cold War examples, deserves to be read by all who want to plan research aimed toward solving real-world problems. In the article, he points out that many environmental problems, as well as many "policy analysis" problems generally, are "squishy"—that is, not readily rendered into mathematical form.

He describes the process of taking a mathematically described model as a surrogate for reality—that is, for a real problem (most often a problem of choice)—then solving the model by logic or mathematical procedure (say computer-based simulation) and using the result as a solution for the real problem. He points out that this method works quite well in the physical sciences but not always very well in solving complex policy problems. The reason is that it is hard to make mathematical models approximate policy reality, where well-understood physical systems are mixed with less well understood biological and social ones. Therefore, it can be much less certain that the solution based on the model is a solution to the real problem. He advocates using mathematical models and solutions based on them as "perspective" on complex problems, to be used "as an aid to careful and considered human judgment. When we forget that and focus our attention too strongly on our methods and our computations as the source of our answers, we not only fail to realize that potential [value of quantitative methodology] but we run the risk of being seriously misled" (Strauch 1975: 12). In this phrase, he prefigures and justifies the field of decision support systems, in which careful analysis is seen to support, but not generate, good decisions.

The broader implication for research planning is that it is extremely rare for the analytical results of one study to fully solve, or inform, a complex problem. Solutions tend to emerge from the complex interplay of research-derived information, field application, and generous amounts of human communication and judgment. This in no way invalidates the use of mathematics or other rigorous science-derived information and procedures. Rather, it means they must be used with maximum information about the context and potential "squishiness" of the problem being addressed.

Another important criterion in problem definition and selection can be called "problem durability" in relation to the timing and timeliness of research. The decision-maker model calls for effective communication between research people and the decision-makers if the latter are in the management or policy world. If the decision that is most important to the decision-maker cannot be helped by research information before the decision must be made, it is not a researchable problem. Thus the projected time the research will take to complete must be matched to the time available to the decision-maker. Usually, this can be accomplished only with detailed knowledge of the problem and the specific research proposed to help solve it. It is to be expected that some applied research projects will stop at this point—that is, when it is realized that they can't be completed in time to assist the decision. This at first seems wasteful, but it is a much greater waste to try to apply expensive research results to a problem that has gone away. Thus, estimating the generality (how many people have the problem and the aggregate cost of wrong decisions taken in the absence of the research) and durability of research opportunities becomes a critical research planning skill.

WRITING OBJECTIVES

Usually, the objective statement, along with the hypotheses formulated in response to the questions posed in the problem statement, will form a separate section of a written plan. An informed reader should be able to judge a plan interesting and doable just by reading the full description of the objectives. This section defines precisely the information of interest, and therefore the data collection tasks to be described in the study design. Objectives may take the form of specific questions to be answered, hypotheses to be tested, or relationships to be quantified or otherwise described. But by the time the objectives are completely written, usually by an iterative process that includes review by others besides the author, it should be clear:

- Whether the results will be qualitative or quantitative.

- Which particular measures of validity are required (statistical analysis, model verification, the meeting of regulatory criteria, or others).

- Whether the data collected will be used for analysis or simple prediction (see Chapter 6).

- What the scope of the study is, in terms of the population to which the results are expected to apply.

Once a problem is clearly defined, the next planning step is to specify the researcher's objective. Students using the decision-maker model to define their problem often have trouble at first differentiating between the decision-maker's objective and their own. The research planner's generic objective is to help the decision-maker reach his or hers by making a correct choice among alternatives more likely. Thus, the research objective may seem at first glance to be only marginally related to that of the decision-maker. For the Park manager with the elk and vegetation problem, the most important ingredient for a solution may lie in a better understanding of the specific adverse effects the elk seem to have on the vegetation. Thus, the researcher's objective might be to define the quantitative relation between elk browsing and the reproduction of the plants involved.

Most often, as indicated for "squishy" problems above, several studies, past and present, will need to be synthesized to clarify the choices for the decision-maker, and, almost always, these will need to be supplemented generously with the decision-maker's experience and judgment. "One problem, one study, one solution" would be nice, but doesn't usually happen.

Research objectives derive from the problem statement, whether using the gap-in-knowledge or decision-maker approach, but they are not congruent with it. In the study objectives, the researcher envisions as closely as possible what will be known when the study is complete. This, of course, is an exercise in forecasting, and, as such, subject to all the uncertainty of all predictions of the behavior of complex systems. Another tongue-in-cheek but useful piece of advice once given to a green researcher was, "Don't title your painting until you are done." This wry comment contains the truth that you often know more, or less, or different things at the end of a study than you thought you would at the beginning. This seems to contradict or at least call into question the setting of specific objectives and to play into the hands of those who confuse freedom of inquiry with lack of goals. This very uncertainty, however, makes it even more important to know at the outset exactly what you are trying to find out, so that you know at the end how reality compares with conjecture, and you therefore can try to explain the difference. The written objective describes what the researcher and the audience expect and is a "risky prediction" à la Popper.

Thus, a good written research objective, at a minimum:

- Clearly describes the intended outcome of the research. If a hypothesis is to be tested the objective section of the plan should state the hypothesis and any alternative hypotheses (see Chapter 5) and should indicate what could be concluded if they were accepted or rejected.

- States clear criteria for assessing whether the objective is accomplished.

- Gives an estimate of the time and money needed to achieve the objective.

- Contains a minimal statement of method, to be expanded later in the plan.

- Is as unitary and specific as possible. A research project often, perhaps usually, has multiple objectives, but the planner should strive to bring them together in one overarching objective directly related to the problem statement. At the least, multiple objectives should be stated in the order of their importance.

- Is realistic enough so that a prepared reader can judge its likelihood of achievement.

TITLES

A title is a condensed objective. It should tell as much and as specifically as space will allow what the study is about and what the author wants to learn. Perhaps, in a practical sense, the most important part of a study plan is its title. Among its many functions, the title is paramount in determining whether a given reader reads on. If the title is confusing or off-putting, the reader goes on to another of the thousands of competitors for her or his attention. The reviewer of a plan wants to know as quickly as possible what the researcher intends to do and why. Titles are also:

- An internal test of whether the author really knows what he or she wants to do. If it proves difficult to write a title shorter than a paragraph, the research problem is not sufficiently defined.

- An indexing device, and as such, should include important key words as signals to potential readers.

It is a good exercise to begin to write titles at the early stages of research planning and to continue to revise them until all elements of the plan are in place. Even though, as noted above, the actual outcome of the research may eventually require titular changes, the challenge of stating the intention of the proposed research in a sentence or less is an important way to focus on the essence of the study. Titles should be as long as they need to be but no longer. Usually early versions are long; titles tend to decrease in length (and increase in specificity) with time and work. Also, be sure to write with the intended reader constantly in mind, and make your title as active as possible (Does it contain a verb? An active verb?). Finally, strive for specificity.

Consider these titles proposed for the same work:

> "Some observations on the demise of selected species of rodents in relation to negative stimuli."
> "Induction of sudden mortality in *Rattus norwegicus L.*: A paradigm for extinction."
> "How to kill rats with a stick."
> "Stick-struck rats die quick."

Each of these fanciful titles has good and bad points: the first is vague but inclusive; the second is unnecessarily stilted but contains an important key phrase, the scientific binomial for the rat in question; the third is concise and clear but does not really say anything about context or science; and the fourth is snappy but also sappy.

The art of writing titles is important and requires a lot of work. Because titles are short, at least relative to the rest of a plan or proposal, they are often given short shrift. That is always a mistake.

FIVE

Creating Hypotheses and Models

Most researchers agree that the most creative part of the scientific process is hypothesis formation. Mechanistic hypotheses make predictions that can be tested and that contain an explanation of why the predicted event is expected. Good hypotheses are "risky predictions," in Popper's term. Often, hypotheses are models that express the most important attributes of a given problem. Models can take many forms, and they can help the research planner in several ways, including the conceptualizing of experiments, organizing and tracking data, and communicating with reviewers. Models that are predominantly analytic or simulative in structure can be used to form and test hypotheses. Strong inference, the use of problem-focused logical trees to construct parallel experiments to test two or more hypotheses, is a powerful tool in planning research; more than one hypothesis should be formulated in answer to a single research question. With carefully constructed hypotheses, it is possible to design novel and good experiments. Without them it usually isn't.

MECHANISTIC HYPOTHESES

In Chapter 2, mechanistic hypotheses were called the "soul" of science. C. H. Waddington, in his book *Tools for Thought* (1977), has this to say about hypotheses and the scientific method:

The essential point about the effective use of the scientific method is not that you try to prove or disprove hypotheses, . . . it is whether one can discover how to ask important questions, and, in relation to each question, can devise experiments which give clear-cut answers one way or the other. . . . It is at least as rare to come across a scientist who consistently applies the true scientific method as it is to find a writer with an impeccably lucid prose style.

Hypothesis formulation, devising questions that can be answered by observation and experiment, is the central creative act of science. Even though some research proceeds without formally stated hypotheses, none is done without at least implied important questions. For purposes of good research planning, attempting to form testable hypotheses has several virtues:

- It forces the planner to think in terms of the "risky predictions."

- It forces the planner to think of ways to disprove what she feels is true.

- It makes explicit the overarching objective of the research (that is, to test an explicit hypothesis).

- It allows reviewers to focus on the core creative act of the proposed research.

In a general sense, any guess about the answer to a question is a hypothesis. To be useful in planning research, however, it must have several additional properties. Foremost, it must be capable of being shown to be wrong if it is wrong. Second, it must contain not only a prediction but also some strong indication of why and how the prediction will be borne out. Most usually, the why and how will come through explanation of an expected pathway of events leading to the outcome predicted.

This latter feature is what differentiates between a mechanistic and a non-mechanistic hypothesis. "Mechanistic" here does not mean "mechanical" in the sense of automatic or predetermined. Rather, it means that the mechanism (the elements and events involved and the sequence of their actions and interactions) by which the outcome is reached is described as fully as possible. Specifying the mechanism has three important functions. First, it serves to convince the researcher and reviewers that the system under investigation is understood. Second, it provides a source of additional tests of the hypothesis (the actions of the elements in the pathway). Third, it helps the later interpretation of results if the outcome is other than the one predicted.

There has been much speculation about where creative hypotheses come from. Some, including Popper, find no structured method; rather, scientists "leap to conclusions" and make risky predictions based on them. Others appeal to "genius," "insight," and extraordinary, non-science experience, including dreams. Peter Caws (1969), however, argues that discovery, in the sense of the creation of new hypotheses, arises by thought processes similar to those used in testing hypotheses, and that this process can be learned. His ideas converge with those of Boulding, as he states: "In the creative process, as in the process of demonstration, science has no special logic but shares the structure of human thought in general, and thought proceeds, in creation and demonstration, according to perfectly intelligible principles. . . . The important thing to realize is that invention is, in its strictest sense, as familiar a process as argument, no more and no less mysterious. Once we get this into our heads, scientific creativity will have been won back from the mystery-mongers" (1380).

This is probably right and good news for the beginning scientific researcher. It means that both hypothesis formulation and testing are learned skills subject to improvement with practice. Constructing chains of logic (arguments) that link data and observation to new testable guesses about the consequences of further experiment or observation is the process to practice. One of the best mind exercises to aid this sort of practice is to try to think how seemingly dissimilar observations, systems, and events might in fact be connected. Jaques Barzun said that "relevance is a property of the mind." By this, he did not simply mean that in some vague sense everything is related to everything else, but that by thoughtful work, connections previously unknown can be found and described.

STRONG INFERENCE

More than forty years ago, John Platt (1964) observed that some approaches to science appeared to work better than others. He observed that "an accumulative method of inductive inference . . . is so effective that I think it should be given the name of 'strong inference.'" This "strong inference" approach was characteristic of rapidly advancing scientific fields, and it consisted of applying the following steps to every problem in science, formally and explicitly and regularly:

- Devising alternative hypotheses;

- Devising a crucial experiment (or several of them), with alternative possible outcomes, each of which, will, as nearly as possible, exclude one or more of these hypotheses;

- Carrying out the experiment so as to get a clean result;

- Recycling the procedure, making subhypotheses to refine the possibilities that remain.

Platt anticipated the criticism that these are the steps science has followed throughout its development, and thus they deserve no name other than the "scientific method." His particular insight echoed that of Waddington; that is, that scientists and even whole areas of science do not make sufficient use of the methods of science. He was concerned that researchers "become method oriented rather than problem oriented." He was sensitive to the additional charge that the strong inference approach concentrated science on systems that were too simplified to mean much in the real world, to which his reply was, "you must study the simplest system you think has the properties you are interested in." This is excellent advice for research planners. He also said, "It pays to have a top-notch group debate every experiment ahead of time; and the habit spreads throughout the field" (1964: 351). This debate is greatly facilitated by the use of written study plans.

Often, the strong inference approach is regarded as an attack by a reductionist scientific elite working with simple systems and a strong quantitative theoretical base on those who study "complex" systems such as cell biology, ecology, social science, and applied sciences focused on environmental and natural resource problems. There is indeed a tendency for the "hard" natural scientists to look down, or at least with some reservation, at other research efforts. This tends to produce not only a surly response from those so regarded but a deliberate avoidance of their methods. That should not be allowed to happen. All should take as much advantage of "hard" methods as possible when planning a specific research effort. It is not always possible to construct the simple, elegant, and informative logical and experimental tree that Platt advocates. For a thoughtful statement of the case that this is so, see Quinn and Dunham (1983). But it is possible to try and to benefit through trying from additional productive thought before beginning to experiment and observe. As H. L. Mencken said about the communists, the best time to argue with them is not when they are right.

MULTIPLE WORKING HYPOTHESES

Over a hundred years ago, Thomas Chamberlain, a geologist (and university president), described the "The Method of Multiple Working Hypotheses" in *Science* magazine (Chamberlain 1965). This is one of the primary ingredients of the strong inference method. Chamberlain made the case that the researchers should force themselves to construct more than one hypothesis for any intended study. This would broaden their view of the problem and prevent them from fixing too soon on a single explanation. The danger to be avoided was the "looking for confirmation rather than refutation" pitfall later described by Popper. Multiple hypotheses kept the researcher in a more objective state of mind. Also, if two hypotheses could be constructed so that the same experiment or observation must invalidate one of them, progress is more rapid, because a valid line of reasoning survives in the remaining hypothesis. Taking a leaf from his own book, Chamberlain also saw drawbacks to the method. He admitted that it could lead to "vacillation," a lack of decisiveness in experiment and conclusion. He also saw problems in teaching the method to the young, who, he said, want simple rather than complex explanations and processes. But it is important in research planning to try to devise multiple hypotheses, to compare them for the risk implied by their predictions and to envision experimental tests that would invalidate them.

MODELS IN RESEARCH PLANNING

Models are simply structured abstractions, constructed to isolate the important elements or components of a system in relation to a particular problem, hypothesis, or theory. They are structured so that they can be used analytically, most often in the test of a mechanistic hypothesis. Because they are, by definition, abstractions intended to choose the most important attributes of the object of investigation so that others may be for the time ignored or treated as "boundary conditions," they are never complete or congruent with the object itself.

Models used in planning and doing science are usually mathematically based, at least in some sense. Verbal or graphic descriptions and physical constructions ("analog" models), however, can also serve as models. For

Table 5.1. A summary classification of models
useful in research planning.

Model attributes	Spatial system (S)	Conceptual system (C)
Analytical (A)	AS	AC
Simulation (S)	SS	SC

purposes here, the vast world of actual and possible models used in environmental and natural resource research are reduced to a gross classification into a four-cell matrix (Table 5.1).

In this classification, a *spatial system* is one with physical boundaries, and a conceptual system is one in which mental boundaries are the dominant descriptor. Thus a spatial system could be an ecosystem, a watershed, or the contents of a laboratory flask. A *conceptual system* could be a representation of the mechanics of the flight of birds, or the functioning of human economic systems. Very often, one will be nested in the other, as in, for example, an economic model of the behavior of small firms in the United States.

An analytical structure most often presents a set of equations that describes a system, the solution of which specifies the outcome of a set of initial conditions. A simulation structure seeks to describe a system, numerically or otherwise, so that its behavior can be predicted, without necessary reference to the structure that makes the prediction. These also can be nested. Analytic models of environmental and natural resource problems and systems derive power and even beauty from mathematics, although most researchers concede that ecological and social processes are not as amenable to mathematical description as more purely physical phenomena. Simulation models usually incorporate values for constants and variables actually observed in the system itself, rather than predicting and analyzing major changes in the system. Most models used in environmental and natural resource research contain elements of both analysis and simulation, but usually one approach predominates. Any given model should present its assumptions with clarity and make its processes as transparent as possible.

Other things being equal, for research purposes, analytical structures are preferred over simulation models. This is because models are often presented as hypotheses in the planning stage, and they thus should have the characteristics of a mechanistic hypothesis. "Modeling" is really a blanket term for approaching a variety of functions important in research planning, including:

- Conceptualizing and visualizing the structure and potential results of experiments

- Organizing and keeping data

- Understanding and communicating concepts and relationships

- Predicting experimental outcomes

- Optimizing inputs for a desired output or set of them

Some think it important to differentiate between "experimental scientists" and "modelers." In reality, researchers who check their models against reality are experimentalists, and any experimentalist who forms a prior idea of what the structure of her experiment will bring is a modeler. Modeling complex processes has become much easier with the increasingly facile use of computers. But the central base of modeling, as of all research, is careful, logical thought. For a basic overview of the thought processes involved in constructing formal models, Hall and Day (1977) is a good place to start.

HYPOTHESIS TESTING

Hypothesis testing, that is, putting the risky prediction at risk, is the opposite of seeking confirmation. Although most researchers desperately want their hypotheses to be correct, the scientific method demands rigorous attempts to prove them false. This next part of the planning process, and of science, is the part arguably most familiar to beginning researchers and non-scientists. It involves constructing or finding experiments that are calculated to deliver a negative result if one is possible (see Chapter 6). The vast literature of formal experimental design is available to help in this, as are all the laboratory and field protocols and instruments that thought and technology have produced. When planning a study, you need to link prior thought about the problem, the questions it engenders, and the hypotheses about answers with a structured experience that includes the design, measurements and observations, protocols, and analysis necessary to draw valid conclusions. The next chapter provides an overview of these and directs the reader to more detailed sources.

SIX

Designing Experiments

Designing experiments is both art and science. Once a hypothesis is formed, there usually exist many possible tests. Choosing an approach that fulfills the requirement of a rigorous test and at the same time is ethically sound and economical in time and materials is difficult. The actual description of an experiment as it is meant to be carried out thus contains elements of the sublime (the intellectual quality of the test) and the worldly (how much it will cost, how long it will take, and who will do it). Choosing the right methods and venue to best test the hypothesis includes making both large and small choices. These include deciding on a quantitative or qualitative approach, laboratory or field (or both) as a venue, using a highly manipulative course or finding an experiment in nature, and constructing a statistical or other model to analyze results. A good written plan documents these and other choices made at the outset and explains why the choices were made. The plan also gives detailed information on the activities scheduled to carry out the research, and the money, people, equipment, supplies, and services that will be needed to complete it. Attention to these details early builds credibility with reviewers and increases the odds of ultimate success (in both getting funds and completing the research).

ELEMENTS OF EXPERIMENTAL DESIGN

The techniques for experimental design have evolved greatly since the early days at Rothamstead, the pioneering agricultural research station in England in the nineteenth century, but the short list of advice published in the middle of the nineteenth century based on experience there is still worth contemplation by environmental researchers (Johnston 1849: 11):

- Everything should be done by weight and measure.

- Both the chemical composition and physical qualities or condition of all substances used should be accurately ascertained and recorded.

- Two experiments of the same kind, one to check the other, should always be made.

- In field experiments the two plots devoted to the same treatment should be as far removed from each other as convenient.

- Land experimented upon ought to be in a uniform, natural, and well-understood condition.

- A second experiment must not be made on the same spot until several years have elapsed.

- All experiments ought to be contrived and executed with a definite object.

- All experiments must be comparative.

Later, agricultural researchers (Little and Hills 1978: 3) gave this list of "steps in experimentation" that still constitute a useful basic checklist.

STEPS IN EXPERIMENTATION

The selection of a procedure for research depends, to a large extent, on the subject matter of the research and its objectives. The research might be descriptive and involve a sampling survey, or it might involve a controlled experiment or series of experiments. When an experiment is involved a number of considerations should be carefully thought through if it is to be a success:

Definition of the problem. The first step in problem solving is to state the problem clearly and concisely. If the problem cannot be defined, there is little chance of it ever being solved. Once the problem is understood, you should be able to formulate questions which, when answered, will lead to solutions.

Statement of objectives. This may be in the form of questions to be answered, the hypothesis to be tested, or the effects to be estimated. Write your objectives in precise terms to allow yourself to plan the experimental procedures more effectively. When there is more than one objective, list them in order of importance, as this might have a bearing on the experimental design. In stating objectives, do not be vague or too ambitious.

Selection of treatments. The success of the experiment rests on the careful selection of treatments, whose evaluation will answer the questions posed.

Selection of experimental material. In selecting experimental material (those organisms and/or objects to which experimental treatments will be applied), you must consider the objectives of the experiment and the population about which inferences are to be made. The material used should be representative of the population on which the treatments will be tested.

Selection of experimental design. Here again a consideration of objectives is important, but a general rule is to choose the simplest design that is likely to provide the precision you require.

Selection of the unit for observation and the number of replications. For example, in field experiments with plants, this means deciding on the size and shape of field plots. In experiments with animals, this means deciding on the number of animals to consider as an experimental unit. Experience from other similar experiments is invaluable in making these decisions. Choose both plot size and the number of replications to produce the required precision of treatment estimate.

Control of the effects of the adjacent units on each other. This is usually accomplished through the use of border rows or buffer strips surrounding treatments and by randomization of treatments.

Data to be collected. The data collected should properly evaluate treatment effects in line with the objectives of the experiment. In addition, you should consider collection of data that will explain why the treatments lead to certain results.

Outlining statistical analysis and summarization of expected results. Write out the sources of variation and associated degrees of freedom in the analysis of variance. Include the various F (or other) tests of statistical significance you may have planned. Consider how the results might be used and prepare possible summary tables or graphs that show the effects you expect. Compare these expected results to the objectives of your experiment to see if the experiment will give the answers you are looking for.

At this point it is well to provide for a review of your plans by a statistician and by one or more of your colleagues. A review by others may bring out points you have overlooked. Certain alterations or adjustments may greatly enrich your experiment and make it possible to learn considerably more from the work you are about to undertake.

Conducting the experiment. Use procedures that are free from personal biases. Make use of the experimental design in collecting data so that differences among individuals or differences associated with order of collection can be removed from experimental error. Avoid fatigue in collecting data. Immediately recheck observations that seem out of line. Organize the collection of your data to facilitate analysis and to avoid errors in copying. If it is necessary to copy data, check the copied figures against the originals immediately.

Analyzing data and interpreting results. Analyze all data as planned and interpret the results in light of the experimental conditions, the hypothesis tested, and the relation of the results to facts previously established. Remember that statistics do not prove anything and that there is always a probability that your conclusions are wrong. Therefore, consider the consequences of making an incorrect decision. Do not jump to a conclusion, even though it is statistically significant, if the conclusion appears out of line with previously established facts. In such a case, investigate the matter further.

Preparation of a complete, readable, and correct report of the research. There is no such thing as a negative result. If the null hypothesis is not rejected, it is positive evidence that there may be no real differences among the treatments tested. Again, check with your colleagues and provide for review of your conclusions.

Formal statistical experimental design and analysis are key ingredients of most research plans on environmental and natural resource subjects. There are many books on formal experimental design, and there is no point in reproducing their ideas here. One of the most accessible books on the subject, regardless of the reader's level of knowledge of the field of statistics, is old but still informative. Sir Ronald Fisher (one of the progenitors of modern statistical analysis of experiments) wrote *The Design of Experiments* more than sixty years ago (Fisher 1935). It still does a good job of grounding a new researcher in the basics of experimental design as seen from a statistical point of view. For a more modern text on the basics of formal experimental design, Cochrane and Cox (1957) is still recommended.

The basic criterion of a good experimental design is whether it results in achieving the experimental objective, which is often to test a hypothesis. Almost all designs incorporate the concepts listed below.

1. *Treatment:* This is something the researcher causes to happen that enhances the chances of achieving his or her objective. For example, if runoff from nitrogen fertilization is expected to cause observed water contamination, a treatment might be applied that prevented runoff from fertilized areas, or runoff might be prevented from entering the observed water.

2. *Control:* In many experiments a control is a no-treatment option. If a pesticide is expected to have a negative effect on species other than its target, a treatment that includes everything about application of pesticide except the pesticide itself might be used as a control. In any event the control represents the usual condition of the system, and the treatment represents the experimentally manipulated condition.

3. *Experimental unit:* This refers to the individual entities to which treatments are applied. Each experimental unit is exposed to some treatment, including the control condition if one is employed. Together the experimental units compose the experimental population.

4. *Replication:* This is the repetition of each treatment two or more times within the same experiment. Replicating treatments allows for informal or formal estimates of the variation associated with the interaction of the treatment and the experimental unit. Because all physical entities measurable in the ordinary world are at some level and degree different, and treatments are used to expose differences due to treatment, the size of a difference that can be detected is determined by the amount of replication. The smaller the difference to be detected, and the larger variability inherent in the experimen-

tal population, the greater the number of replications required for a valid test of a hypothesis about difference among treatments.

5. *Randomization:* This is the assignment of treatments to experimental units so that all units in the experiment have the same chance of receiving any of the treatments included. If only some of the experimental units can receive one treatment, and a second group can only receive another treatment, there is no way to know whether inherent differences in the groups of experimental units, rather than the treatments, caused observed differences in outcomes.

6. *Experimental error:* This term refers to the calculation of the probability that the values estimated for the various treatments are actually different. It depends on the presence of replication, assumptions about variation in the true population from which experimental units are drawn, and on the variability of the values obtained from experimental units within and among treatments.

Regardless of the field of study, or the kind of experimental design being contemplated, you should consider each of these concepts in designing a study. Reduce any uncertainty about them by consulting the appropriate book, such as one of those mentioned above and at the end of the chapter, and, preferably, a statistician, in person or online. Make use of a professional statistician, a valuable reviewer of any research plan, including those rare plans that make little use of formal statistical methods.

The methods section of the plan should contain these ingredients, each fully explained:

- The equations and narrative that describe the experimental design.

- The statistical and other tests, along with the analytical procedures that will generate the tests (as indicated by the design).

- A list of variables and factors (sources of variation) to be considered in the experiment, with the variables to be used to describe them. Factors are general attributes of individuals or systems (tree size, water quality), and variables are the measured quantities used in the experiment to represent them (tree volume in cubic meters, nitrate content of water in moles per liter).

- A description of which factors will be represented by experimental variables, which are sources of variation to be controlled in the experimental design, and which are uncontrolled sources of variation.

- Graphs or other representations of what the results of the experiment are expected to look like for each hypothesis and set of variables previously described.

- Laboratory and field procedures in sufficient detail (including references) so that another investigator in the same field could carry out the study. For field studies, accurate maps of experimental locations are a must.

- Computational procedures and the necessary hardware and software to carry them out.

- A list of necessary permissions and regulatory rules that must be obtained or met to carry out the research (human subjects clearance, property access, controlled substance use, radioactive material handling and disposal, for example).

- An assessment of the environmental impact of the experimental manipulations, if any.

The ultimate test of a good methods section is completeness. Not only does this reassure the current researcher that the study is doable, it provides for continuity in case another has to complete the study.

FINDING EXPERIMENTS

Jared Diamond (1986) described three classes of experiments useful to community ecologists: laboratory experiments (LE), field experiments (FE), and natural experiments (NE). In his terminology, LEs and FEs have imposed treatments and environmental manipulations, and NEs are "found" experiments. He pointed out that the "historical" sciences, those that observe events more than they manipulate them (geology, astronomy, and others), rely on natural experiments for most of their hypothesis tests. He also treated the three kinds of experiment as a continuum. For example, NEs must be manipulated at least to the extent that they are observed or measured. He separated NEs into two kinds: natural trajectory experiments (NTEs), which are comparisons of the same community before, during, and after some disturbance event not caused by the experimenter, and natural snapshot experiments (NSEs), which are comparisons of communities assumed to have reached a steady state or something approximating it. This classification seems useful for most experiments on environmental and natural resource problems, and you should consider their strengths and weaknesses carefully in light of your objectives.

In essence, "found" experiments (NEs in Diamond's terminology) gain in generality (the number of communities or places to which the results clearly apply) at the sacrifice of control, the hallmark of laboratory and constructed field experiments. Finding good experiments demands familiarity with the systems studied and an intimate knowledge of sampling theory and practice. Devising logical and physical linkages among laboratory, field, and found experiments is both a good mental exercise and a way to ameliorate the drawbacks of each experimental type.

RESEARCH AND ADAPTIVE MANAGEMENT

Adaptive management is the term applied to management activities that have knowledge as one of their formal goals (Bormann et al. 1996). This involves constructing experiments within a management framework or activity, such as a timber harvest, a watershed vegetation restoration, or a pollution abatement project. In its simplest form, it blends experimental design principles with management protocols in ways that do not compromise management goals and that result in credible information. It almost always costs more to manage adaptively, in this sense, and therefore adaptive management can be done only when the information gained is expected to more than offset its cost. One of the richest sources of information of this kind is through "found" retrospective experiments (a subset of the natural experiments described by Diamond), in which different management prescriptions have been applied to similar systems. But because of the lack of control of the activities at the outset, the experiments have to be carefully crafted to be credible. That means following the steps rigorously: forming hypotheses, devising the test, sampling the systems to conduct the test, and drawing conclusions with a statement of their probability of being due to chance.

QUALITATIVE METHODS

For some problems in environmental and natural resource research, it is impossible to make quantitative measurements of the variables of interest. For example, the opinion of people living in and around forests may be important to the choice of forest management and protection methods. But, beyond a statement of how many, or what percentage of them, approve or disapprove of a given practice, it may not be possible to quantify their views. In this instance, the construction of a narrative vision for what they would like the forest to look like and do might be the object of research and an important element in the solution of the problem of choosing among management techniques. A rich literature on qualitative techniques is available for consultation. Every planner of environment and natural resource research, no matter how "quantitative" their research is, should be familiar with the basic techniques of qualitative research. Almost every such study raises qualitative questions, and qualitative methods (most obviously review of the literature) are often important methods to use to define a problem and its context. The appendix at the end of this chapter suggests some useful qualitative methods readings (S. T. Warren, pers. comm., 1991).

LIMITATIONS OF RESEARCH

Scientific research is subject to at least three kinds of limitations: ethical, methodological, and structural. Ethical limitations range from consideration of what it is permissible to do and not do to research subjects (especially humans, but also plants and animals in the view of some) to cosmic questions such as whether there should be self-imposed limits on human knowledge. Each scientific study is planned in a matrix of law, culture, custom, and individual beliefs that constrain it. A careful thinking through of these limits and how they may apply to the planned study is an important ingredient in the preparation of a written study plan. The most important component of ethical review is detailed self-knowledge (Thomas 1993), but all applicable rules (for example, those regarding experimentation with human subjects) must be reviewed and accepted.

Methodological limits include the time lag from research plan to result, the uncertainty principle that decrees that measuring a system may change its properties in unknown or unknowable ways, and simple physical limitations on what can be measured. A methodological limit that is perhaps too often ignored is objectivity of problem definition and choice. A problem that is incorrectly or ineptly defined usually does not lead to a useful result.

Structural limitations refer to limits imposed by logic and its mathematical and verbal expression. It has been noted above that scientific method is essentially silent on concepts like "eternal truth." Research only proves things wrong; that which is designated as "right" is right only by default. It hasn't yet been proven wrong. Correlation, the observation that two or more variables move together in some predictable way, does not constitute proof, or even strong evidence, that one causes the other. Nor does order in time establish causality. *Post hoc, propter hoc,* "after this therefore because of this," is a logical fault. Only if the events can be connected by a plausible mechanism can causality be inferred. Looking for confirmation, according to Popper, above, does not test or strengthen a hypothesis; only a rigorous attempt to prove it false that fails does that. These structural limitations of science are mostly well known, but you should examine your written plan to make sure it doesn't incorporate any variations of these grand flaws.

All studies are bounded not only by the creativity, technical capability, and diligence of the researcher but also by time and money. Thus the design of experiments (here used in its broadest sense to mean any purposeful attempt to meet a research objective or test a hypothesis) involves the interplay between the desirable and the possible. Indeed, one of the most useful outcomes of written study plans is to show when a proposed study, one that is valid in terms of the problem addressed and the questions asked, is beyond the financial or technical means of the researcher. Thus, the methods portion of a written plan necessarily includes not only the study design, procedures, and analysis but detailed estimates of the time and money it will take to carry them out.

Table 6.1. A rudimentary two-year, three-column budget.

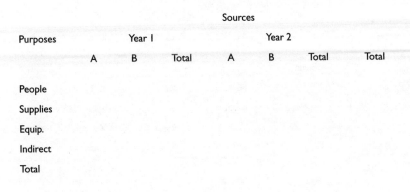

Purposes	Sources						
	Year 1			Year 2			
	A	B	Total	A	B	Total	Total
People							
Supplies							
Equip.							
Indirect							
Total							

THE THREE-COLUMN BUDGET

Any worthwhile research plan includes a careful estimate of how much the research will cost and who will pay. This estimate is best portrayed in what is commonly called a three-column budget (Table 6.1). In this format, items to be paid for are presented in the left-hand column, in sets of rows that indicate the purposes to which funds will be allocated, such as salaries, equipment, supplies, and organizational overhead. In the columns to the right of the purposes, each period of time over which the research will take place (often, but not necessarily, years) is represented by a column for each source of funds. An additional column for each period represents the total funds allocated to a specific purpose during that time. A final row and column presents the total for each source and time and for each purpose. Budgets of this kind quickly inform the planner and reviewers exactly what resources are needed and when, and from whom they will be sought. The tables are technically easy to construct using computer spreadsheets. Their virtue lies in the hard thought about sources and uses of funds that they cause.

Table 6.2. A rudimentary five-task, five-month schedule.

Task			Time, Months		
	1	2	3	4	5
1	Begin	X	End		
2		Begin	X	X	End
3		Begin	X	X	End
4			Begin	X	End
5				Begin	End

PORTRAYING SCHEDULES

Perhaps the principal practical virtue of a written research plan is to get a specific idea of what needs to be done when to carry the research to a useful conclusion. The easiest way to portray this in a written plan is with another spreadsheet-based table (Table 6.2). In this one, the planned research is broken down into detailed individual tasks, and these are listed in a left-hand column in the order of their inception (date of beginning). Time periods of the study are listed in columns to the right in units useful for the planned study (days, months, years), and a line is drawn in the row next to each task that begins at its planned start date and ends at its completion date. Special dates (reports, deadlines) are noted as well. This picture quickly shows periods of maximum activity (many lines vertically overlapping) and indicates when the researcher and other participants will be heavily or lightly engaged. Very detailed and complicated planning and scheduling software is available, but it is a good idea to start, at least, with this simple portrayal.

In sum, the plan must give a concise picture of how much money will be needed to complete the project, what it will be spent on, and how the time allocated to the project will be spent.

APPENDIX: BOOKS ON USEFUL QUALITATIVE METHODS

Bristin, R. W. 1973. *Cross-Cultural Research Methods.* New York: Wiley.

Casley, D. J., and K. Kumar. 1988. *The Collection, Analysis, and Use of Monitoring and Evaluating Data.* Baltimore: World Bank/Johns Hopkins University Press.

Casley, D. J., and D. A. Lury. 1981. *Data Collection in Developing Countries.* Oxford: Clarendon.

George, A. L. 1979. Case studies and theory development: The method of structured, focused comparison. In *Diplomacy: New Approaches in History, Theory and Policy,* P. G. Lauren (ed.). London: Free Press, pp. 43–68.

Grosof, M. S. 1985. *A Research Primer for the Social and Behavioral Sciences.* Orlando: Academic Press.

Heberlein, T. A. 1988. Improving interdisciplinary research: Integrating the social and natural sciences. *Society and Natural Resources* 1(1): 5–16.

Kirk, J., and Miller, M. L. 1985. *Reliability and Validity in Qualitative Research.* Beverly Hills: Sage.

Krueger, R. A. 1988. *Focus Groups: A Practical Guide for Applied Research.* Beverly Hills: Sage.

Lijphart, A. 1975. The comparable-cases strategy in comparative research. *Comp. Pol. Studies* 8(2): 158–177.

Miles, M. B., and A. M. Huberman. 1984. *Qualitative Data Analysis: A Sourcebook of New Methods.* Beverly Hills, Sage.

Morgan, D. L. 1988. *Focus Groups in Qualitative Research.* Beverly Hills: Sage.

Whyte, W. F. 1984. *Learning from the Field: A Guide from Experience.* Beverly Hills: Sage.

Yin, R. K. 1984. *Case Study Research: Design and Methods.* Beverly Hills: Sage.

SEVEN

Communicating Study Plans

If there is a required skill that cuts across all professions, including re-search, it is the ability to communicate effectively. No matter how important a researcher's hypotheses are, they will have no effect on science and the world if no one else understands them. "We are all prisoners in our own skin," as the saying has it. Unless our ideas escape that prison in intel-ligible form, no useful research is done and no problems are solved.

Almost every written research plan that is important enough to draw no-tice is discussed in person as well as read. Thus, most research planners will present their plans to formal or informal audiences by talking about them.

The elements of effective communication are:

- Writing

- Speaking

- Listening

- Reading

- Observing

While the focus here is on written plans, the other four elements are necessary to the overall process. Research plans are always talked about, in formal or informal meetings (see below). Only by listening to reviewers and colleagues and reading their comments can the review process lead to improved research. Observation, not only of the objects of research but of colleagues and the environment in which the research is to be conducted, can greatly improve chances for success.

WRITING CLEARLY

Writing about writing is popular, and there are many good articles and books about science writing. Here is a checklist to remind the research planner of the most basic ingredients of effective communication. Four major principles need to be constantly held in mind and reviewed (Perry 1993):

Know your topic. This is often assumed as a given in research planning, and, indeed, by the time a researcher is planning independent study (say at the outset of doctoral research), she usually has accumulated lots of information. But until a synthesis of information important to your study exists in your own mind, to the point where you are generating new ideas and insights about it, you don't know your topic sufficiently well to write a study plan. Review the literature that covers all the major references to your subject, particularly those over the past decade, as a basis for knowing your topic; it is essential to show in your study plan how the most important and recent papers relate to each other and to your topic. Usually, five or ten key references frame the problem you are describing and bear on the questions you intend to ask. These references should be more thoroughly described and synthesized. Often, this synthesis is written into the body of the study plan, and the larger literature review is included as an appendix. The key to good writing is critical reading, lots of it. Annotating articles in your own words, with comments on your intended use of them, is a good habit to cultivate.

Know your audience. Knowing in a general way who you want to reach with a plan is an important first step, but until you can name and locate at least two actual people who you want to read your plan, you haven't defined an audience. Once you have identified these "real" people, think about them. What interests them? Why? Try to put yourself in their place, and ask, "why would I want to read a plan with the title . . . ?" To merely think, as is often true, that they must read your plan as part of their job (grant reviewer, major professor) is insufficient if you want to extract maximum value from their review.

Know your periphery. Most granting agencies, major laboratories, and many fields of science have their own rules for writing. General style references (for example, Strunk and White 1979 and subsequent editions, Council of Biological Editors 1994), guidelines for proposals available from granting organizations like the National Science Foundation, and institutional writing guidelines and procedures (for example, those of universities and research institutes) are all sources of help in reaching your audience. Make sure you have done a thorough search for and of these resources. Most are available on the Internet.

Know yourself. Treat plan writing as a task at least equivalent in importance to, for example, carrying out laboratory protocols. That means you must prepare carefully, make sure all the needed tools are available (at a minimum, a dictionary), and create an environment that will support your writing effort. Many a would-be writer has passed from college writing courses to Social Security checks while waiting for inspiration. So with the writers of research plans. Learn your own writing habits: when, where, and under what conditions you do your best work. Then create those conditions as nearly as you can and sit down and write. Once you know your topic, your audience, and your periphery, all that remains is to master yourself. That will be much easier if you have done a good job on the first three.

Becoming a good reviewer is a good way to ensure that others will give your ideas and proposals useful review. The *Canadian Journal of Forest*

Science gives concise and practical advice to those who review its articles for publication (see below). The main thing to remember is that you are trying to help the author. A condescending or hectoring tone makes authors not want to take your advice, making it more likely that you have wasted their time and yours.

DESIRABLE QUALITIES OF REVIEWS AND NOTES FOR REVIEWERS (ADAPTED FROM THE *CANADIAN JOURNAL FOR FOREST SCIENCE*).

a. *Objectivity:* Strive to be objective in your appraisal. Objectivity may be difficult since the manuscript you receive is on a subject related to your own interests and since you are probably well acquainted with the author.

b. *Accuracy:* A good procedure to follow is to verify those points you wish to make that take little time. If you are not sure of your grounds, do not hesitate to comment, but write in such a way as to reflect your uncertainty.

c. *Relevance:* Confine your comments to an evaluation of the subject matter in the manuscript. Do not criticize the manuscript for lack of subject matter except as essential for establishing the points that are covered or for encouraging further work.

d. *Courtesy:* Authors are naturally proud of their work and may be sensitive to sarcasm or discourteous remarks. Comments are most likely to be acted upon in the constructive manner you intend if they are made with courtesy.

Notes for Reviewers

1. *Title:* Does it adequately describe the subject of the manuscript? Can you suggest an improvement in wording?

2. *Summary page:* It should concisely relate the background, pur-

pose, objectives, conclusions, and recommendations of the report in client-oriented terms. Special attention should be focused on this portion of the review.

3. *Review of literature:* Is due credit given to relevant contributions of others? Is this contribution in proper perspective in relation to the state of knowledge in the field? Is the number of literature citations appropriate?

4. *Methods:* Were the methods appropriate? Are they described in sufficient detail to permit competent readers to repeat the work; or, are sources cited in which the appropriate detail is given?

5. *Organization:* Is the subject developed logically and effectively?

6. *Duplication:* Is the published work of the author or others repeated unnecessarily? Can the manuscript be shortened without loss of content?

7. *Presentation of data:* Should data presented in tables be given in figures or appendices or vice versa?

8. *Tables and figures:* Are captions, headings, and units adequate to make the table or figure understandable alone? Can layout or design be improved? Are statistical reliabilities shown?

9. *Conclusions and recommendations:* Are they adequate, supported by the data? Do conclusions relate to and follow objectives? Is more research warranted?

10. *Conjecture:* Does the author distinguish clearly between conjecture and fact? Is the amount of conjecture excessive?

11. *Literature cited:* Are there obvious errors like misspelled words or names?

12. *Readability:* Can a well-informed client understand what is being said? Is the report free of unnecessary technical terms and jargon?

13. *Metrication:* Are all units approved by the SI system? Are useful conversions given when needed?

SPEAKING CLEARLY

The best concise speaking advice I have ever received was from an undergraduate speech teacher whose name I have regrettably forgotten: "stand up, speak up, and shut up." All four of the principles above apply to speaking as well as to writing, but effective spoken communication differs importantly in method and content from writing (Bragg 1966). The differences are particularly important when presenting research plans because they are complex and specialized.

Your body language, how and where you look and how you act, strongly affects how and whether your audience gets your message. To American audiences, at least, a relaxed but purposeful manner is usually best received. Nothing beats preparation for allowing you to appear relaxed and purposeful. Practice your talk several times, with an audience if possible, and speak from an outline. Reading a research plan word for word will lose almost any audience.

Remember that once you say something, you can't take it back effectively. You can always edit a written plan after the fact of writing. Once you say something, your audience has it for all time, despite attempts at later repair. This argues for conservatism in oral presentations of research plans. It is fine to be enthusiastic, but don't let enthusiasm carry you far into unsupported speculation. Make sure that the facts you present are such, as well as you can, and don't embroider. Absolutely avoid ad hominem comments. Using your cutting wit to put in their place those who disagree with you about hypothesis or method is often fun but usually counterproductive and certainly is aside from your purpose, which is to get your audience to understand exactly what research you intend to do. Try rephrasing negative comments as positive suggestions for change, either out loud or to yourself, as appropriate.

Speeches need to include at least some, and usually substantial, repetition, because the audience can't "flip back to page three." Building this

73

repetition into a talk without being boring or appearing to talk down to your audience is an acquired skill. No better advice exists on the point that the old admonition to would-be teachers: "tell 'em what you are going to tell 'em, then tell 'em, then tell 'em what you told 'em." A summary of major points at the end of a talk is often its most useful part. The audience has probably absorbed a lot of what you have said but needs a framework to put it on.

The order of presentation is radically different in a spoken, as opposed to a written, presentation of a research plan. In a spoken plan, it is important for the audience to know at the outset exactly where you are going. Often it is effective to present your intended outcome as a proposition to be argued. "I hope to convince you that . . ." is an opening that gets attention and focuses the audience on where you want to go. Researchers often want to spend most of their time on "how to get there"—that is, on detailed renditions of methods. Remember that most humans have an optimum attention span of about twenty minutes, and make sure they know where you want to go with your research, not just what methods you will use. Give them the punch line first. A good order is: introduce, then give conclusions (research objectives and why they are important), then support one or two clear propositions with argument and data (what your method is and why you chose it, for example). Then restate your problem and objectives with a clear indication of the time and money you expect your project to use.

Finally, you can respond directly to indications of lack of understanding, boredom, or hostility on the part of your audience, and you should do so, but not at the expense of losing the thread and momentum of your presentation. If you don't want questions during your talk, make that clear, and make clear that you will leave time for them at the end of your remarks. Always repeat questions from the audience unless someone else does it for you.

The mechanics of effective speaking are deceptively simple. It is in their unerring application that good speakers excel the others. To be a good speaker:

Know your environment and audience. Knowing your audience in advance is critical to your success as an oral presenter. Really work to learn as much as you can as soon as you can.

The environment in which the talk is delivered becomes an integral part of your presentation. If possible, practice your talk in the place you will actually give it. It is best to practice before a live audience, but don't be afraid to practice your talk alone. Microphones, computers, projectors, and light switches become saboteurs as soon as your presentation begins. Master them ahead of time by practicing with them. Know whether your visuals can be seen and interpreted with ease from the back row. If they can't, either revise them so they can be, or don't use them.

Write down your objective in making the talk (not the objective of your research, but why you are presenting it to this audience). Do you want permission or money to do the research? Do you want critical comment on your hypotheses or methods? Are there particularly skeptical views represented in the likely audience? What are you going to do about them? How, at the end of the presentation, will you find out if you achieved your objective?

Make an outline of your presentation, even (perhaps especially) if you have a complete written version. The outline should be in sufficient detail to let you deliver the talk without other support. It should indicate exactly when to show each slide or other visual aid you intend to use.

Make your visual material clear and simple, and explain each slide completely. Confusing graphs, tables with hundreds of tiny numbers, and complicated flow diagrams that might (emphasize *might*) be admissible in written form are the quickest route to a sleepy or even angry audience. Vary your visual material. An endless succession of bullet-pointed blue slides (often with a well-known PowerPoint pattern superimposed) is also an audience killer. In general, people like to look at real things: show a photograph or drawing of what you are talking about in between tables and graphs, and keep text slides to a minimum. Tufte's *Visual Display of Quantitative Information* (1983) is a good reference for preparing materials for spoken as well as written presentations.

Practice your delivery in front of a live audience if possible, then present confidently. Speak clearly, but not too slowly, and allow your natural enthusiasms and emphases to come through in your voice modulation. Be sure your major concepts, conclusions, and particularly your research objective are explained at the outset. Keep eye contact with all parts of the audience, and never talk in the dark. Include sufficient repetition to ensure understanding, and end on time with a restatement of your objective and conclusions. If at all possible, involve your audience directly in the presentation by allowing them to ask questions as you proceed, by asking them questions, or both.

It isn't often possible, but if you can, get written feedback from your audience. Keep a record of what you think went well and poorly for each presentation. If you can, talk with a member of the audience to get his or her specific opinion on your delivery as well as your research plan. In a classroom or rehearsal situation, you could use a simple form to get feedback (see sidebar).

AN ORAL STUDY PLAN PRESENTATION FEEDBACK FORM

How well did you understand (5 = perfectly, 1 = not at all):

> The problem statement: why is the research to be done?
>
> The objective: what will be accomplished?
>
> The procedures: how and where will the research be done?
>
> The scope: how much will it cost and how long will it take?
>
> The results: who will use the results and how will they get them?
>
> The urgency: why do this now?

The presentation: how effectively were language, structure, visuals used?

Give your three most important suggestions for improvement.

Use the feedback you receive to revise your presentation for future use, incorporating what you have learned. This exercise is valuable even if you never talk about the same topic again, but usually you will. "Put your lecture notes on durable paper."

The more presentations you make the better at it you will be, particularly if you follow the advice above.

RUNNING MEETINGS

Research planning and plan presentation and discussion are often done by groups, either formal or informal. Meetings, while ranked as leading time-wasters, are also at least as necessary to the progress of science, and thus it is important to know how to effectively run and participate in them. Heiligmann and Rudolph (1978) give detailed advice on how to make sure that formal meetings are a success. All researchers that go to or run meetings should read it periodically. The basics are easy to list, but hard to put into practice. Good meetings have:

- A single objective, with clear criteria that allow participants to know when it is met.

- A set time that is never exceeded. If the objective hasn't been met at the end of the allotted time, schedule another meeting or find another mechanism for completion.

- Participation by everyone present. Often, one wants to say always, those who naturally say least have the most to contribute. A good meeting manager uses all the human resources in the meeting and doesn't allow prolixity to trump intelligence.

TALKING WITH MANAGERS

Environmental research problems often arise from environmental management problems. If the decision-maker model of problem definition is used, then it is necessary to talk with managers if they are the decision makers in question. C. P. Snow ([1957] 1993) talked about scientists and humanists as "two cultures." Researchers and managers also often seem to come from completely different worlds, for good reasons, and therefore have trouble understanding each other. As with all communication across gaps in understanding, effective listening is key. Recognizing that each person has different skills, values, and environments will go far to improve information exchange. Researchers live by proving things wrong, as previously noted. To most managers, who must be positive advocates for the achievement of difficult, time-constrained objectives, this is a negative and discouraging stance. As the manager of a large natural resource agency once told me (as the manager of a research program paid to help him), "All you bastards ever do is prove me wrong." Researchers, on the other hand, often feel that managers disregard the complexities of science and nature, and therefore err.

As described in Chapter 6, having a clear problem definition to refer to can provide a useful framework for mutual listening. If the researcher focuses on trying to understand the manager's objectives and constraints, and the manager focuses on the researcher's capabilities to provide new information, progress in understanding and problem definition are usually rapid. Too often the researcher wants to criticize management objectives, and the manager wants to say what research to do and how to do it. Thus both tend toward their minimum knowledge set. Researchers may indeed have valid criticisms of a manager's objectives, and managers are often astute critics of research directions and methods. But most discussions proceed more fruitfully when they start with areas of agreement, and mutual criticism works better after mutual respect is established.

Mutual respect is established only after people know enough about each other to make a reasoned judgment. The best route to more effective communication between researchers and managers is extended personal contact. Opportunities for this have to be manufactured against the gradients of daily business. Taking full advantage of occasions to talk that arise in the course of business is thus the first route to understanding. During the informal time at meetings that include both researchers and managers, each sticks to his or her own group. Changing this pattern, perhaps seeking out the company of managers at coffee break or lunch, can pay dividends in improved understanding. Field trips that include both researchers and managers are often fruitful opportunities to talk about issues important to both. The key is frequent, continued contact.

WRITING GRANT PROPOSALS

Writing research proposals to funding agencies is a part of the life of almost every researcher. A good study plan is almost always a good grant proposal if it is in the format and language of the potential funder. Several points about funding proposals should additionally be considered:

Ideas come first. If your statement of why you want the money begins "I want . . ." or "I need . . ." you are probably on the wrong track. The best summary statements of research intent begin "I think . . ." and are followed by a hypothesis. "I think that microbial populations drive plant succession on mixed-forest sites by changing cation exchange capacity."

Match the idea to a money source. Granting agencies, foundations, and individuals have their own goals and agendas. They fund good ideas that help them achieve their own purposes. Begin with a search, using all available help, for sources that have funded ideas like yours and whose literature indicates a continuing interest. Carefully match your idea to their stated goals. Then make this explicit in your study plan.

Prepare a one-page selling document. This is used to make contact with as many sources of funds as possible, always being aware of their rules and those of the institution from which the proposal is being submitted. A good one-pager includes: title, author, institution, objective with hypotheses, problem, methods (very brief), and an estimate of the time and money required for completion.

Contact all the sources you have identified. Use both formal and informal channels. Determine how receptive the source is to a more detailed proposal. Get to know the source and its goals as well as you can by reading and listening.

Prepare a detailed written proposal that rigorously follows the source format and guidelines for submission. Make sure any institutional requirements and protocols on your end are met.

Submit the proposal with a cover letter that specifies to what request for proposals or part of the source organization your proposal is submitted. Also include your proposal title, the names of reviewers (if you have their permission), and to whom the source should reply. If you work in a university or another large organization, a professional "grants and contracts" unit may provide the cover letter for you—but you, as the researcher, should check it.

Follow up. Respond immediately and fully to any request from the source for further information.

Perform. The best assistance in getting future grants is good performance on the ones you have had in the past.

The written study plan is one element in a long-term communication strategy, usually beginning with informal discussions and musings and continuing through the publication and distribution of the results of the research. Seeing the strategy as a whole, and one that will be implemented over a long time, often years, helps one see its importance and the place of the study plan in the strategy more clearly.

EIGHT

Understanding the Role of Science

Science and society interact in important ways. Good practice in environmental professions is based on accepted science. Publics at large depend on science as a basis for law and regulation, and expect science to provide positive benefits across a huge range of human activities. Most basic science is supported directly by public funds, and all science is supported either directly by tax revenues or by consumer expenditures. In return, science is accountable to legislatures, corporations, and charitable foundations. The complex relationship between science and society is not always a comfortable one. To plan research effectively, it is necessary to consider how the intended research fits into the web of interactions among scientists, institutions, and publics.

SCIENCE AND PROFESSIONAL PRACTICE

Scientific people and institutions pursue two broad societal goals that roughly correspond in intent to the "basic" and "applied" research definitions given earlier in the book. The goal of basic research is to understand nature and the human condition in it. Such research is undertaken without the expectation of any practical outcome other than greater understanding of the universe and its contents. History clearly shows, however, that this understanding so gained frequently leads to previously unsuspected practical outcomes. Retrospective studies of science-based innovations that have become societal and commercial successes often find their roots in basic science (Illinois Institute of Technology 1968). Thus, public funding of basic science is often justified by the probability that the knowledge gained will be practically useful as well as enlightening. This bargain between science and society has its modern origin and rationale in the United States best described in a report to the president by Vannevar Bush (Bush [1945] 1990). In the report, Bush says, under the heading "Scientific Progress Is Essential":

> Progress in the war against disease depends upon a flow of new scientific knowledge. New products, new industries, and more jobs require continuous additions to knowledge of the laws of nature, and the application of that knowledge to practical purposes. Similarly, our defense against aggression demands new knowledge so that we can develop new and improved weapons. This essential, new knowledge can be obtained only through basic scientific research.
>
> Science can be effective in the national welfare only as a member of a team, whether the conditions be peace or war. But without scientific progress no amount of achievement in other directions can insure our health, prosperity, and security as a nation in the modern world. (10–11)

These paragraphs display the continuing rationale for government funding of science. The applied goal is to overtly assist others in achieving practical ends—for example, conquest of a disease, provision of clean water or timber, or the introduction of new products. Funding for applied research is usually tied to some reasonably near-term practical goal. Most often, the products of applied research are passed into the hands of professional practitioners such as medical doctors, land managers, or production engineers. In this kind of research, particularly if the decision-maker model of problem definition is used, it is critical for the research planner to understand the practicing professionals to whom the results are to be given.

It is also true that professional practitioners, or, more broadly, "technology," aids scientific research as much as science helps them, or it. Joseph von Fraunhofer, who discovered the optical signatures of the elements, was a German glassblower. Sadi Carnot, the formulator of the second law of thermodynamics, was a French engineer (in the exact sense of the word; he ran steam engines). Open communication between researchers and practitioners helps both, particularly when new ventures are being planned, so it is important for researchers to have sound knowledge of the similarities and differences between the practice of science and the methods and outlook of those whose professions depend on and contribute to science.

Professional practitioners, here environmental and natural resource professionals, use their skills, judgment, and ethical framework to help other people achieve their goals. Professional practice is not an end in itself (Stoltenberg et al. 1970) but rather an outward-looking activity that finds its satisfaction (and pay) in improving the lot of clients. Clients may be individuals or organizations, in the public or private sector, but their objectives rule the activities of the professional, and, within ethical limits, professionals will use their best skill and effort to see that their clients get what they want. Thus, for example, environmental professionals are fundamentally different from environmental advocates. Environmental advocates believe they know what is best for society or the world and work to realize their vision of what is right. They often feel justified by the importance of their vision to use any means at hand to move toward it. Environmental professionals, on the other hand, work to help others achieve their vision and are restricted to the use of a verifiable and credible set of professional skills to do so. Research scientists often see themselves as pursuing their own goals; they work to fill in a personal vision of what most urgently needs to be known. They too must present a verifiable and credible set of skills, but they often resemble advocates more than professionals in that they work toward their own ends. This means that fairly often the notion of helping others achieve their goals is an unfamiliar concept, a major potential barrier in communication between researchers and practitioners. It may also explain why scientists often feel free to become advocates, even outside their field of expertise.

For the reason given above, and for several others, it is imperative for every practitioner to have a good grasp of the nature and processes of science. Professionals are expected to know and interpret to others (clients) the science products found in scientific publications. They are also expected to be able to explain clients' information needs to researchers (this is a primary skill for practitioners in such outreach organizations as the Federal and State Cooperative Extension Service). Education, based in part on the products of science, is often a primary role for practitioners with respect to employees and publics. For all these reasons, practitioners need to remain current in the methods and substance of the scientific research underlying their practice.

GENDER AND SCIENCE

In the late twentieth and early twenty-first centuries, women have been emerging as a much stronger force in the research community than previously. This has caused a flurry of studies to determine the causes, effects, problems, and opportunities associated with this profound change. When I began my research career, shortly after the middle of the twentieth century, there was still a publication called *American Men of Science*, for example, and it purported to be a quite complete listing of scientists in the United States. This trend to greater female participation in research will certainly continue and grow, so it is important for each researcher concerned with the role of science in society to understand its implications.

Most members of the research community agree that greater female participation is a good thing. The roughly 50 percent of the population that had been mostly excluded from science now has a greater opportunity to participate, with the attendant higher probability of increasing the pool of talented people entering scientific careers. Also, as more woman scientists emerge, more female role models will be available to influence the young positively about science.

Problems have been noted, however, and most have not as yet been positively addressed. Women remain the primary care givers for children and organizers for families. These duties often fit poorly with the rhythm of a scientific career and the usual daily routine of researchers. Single-minded, round-the-clock effort and participation are expected of graduate students and many members of scientific teams. Research leaders are expected to give first priority to the pursuit of funds and to presence in the laboratory, library, and field. Those who dilute these activities with "personal" agendas are often viewed with suspicion by colleagues. Do they have the necessary commitment to become "first-rate" researchers?

The long male-dominated history of science also presents problems. When an early-career female publishes with a late-career male with a strong reputation, the publication is subconsciously credited to the male by many.

It has also been suggested that women and men differ in their intellectual approach to scientific research. It appears that women, on average, publish fewer, but higher-quality, papers than men (Sonnert and Holton 1994). This lower productivity (in numbers) may result is a slower rate of advancement for women in systems (for example, tenure processes at universities) that reward rate of publication.

Even though problems remain, the opening of science to women on large scale is an event to be celebrated and encouraged. The more widely the ideas and methods of science are understood and practiced, the stronger the whole scientific edifice will be.

THE ORGANIZATIONAL STRUCTURE OF SCIENCE

Scientific research, in the current world, is done in and by organizations. In the United States, much research is publicly funded, through federal appropriations to the National Institutes of Health, the National Science Foundation, and cabinet departments such as Defense, Energy, Interior, and Agriculture. The latter three fund much of the country's natural resource and environmental research, on topics related to the missions of the departments and their agencies and bureaus. For example, the largest single forestry research budget is that of the research branch of the Forest Service, in the Department of Agriculture. Much applied research and development is thus federally funded, but a substantial amount is funded through private institutions and businesses. Research universities, often in collaboration with government, industry, or nonprofit organizations are major contributors to natural resource and environmental research.

Almost all research is subject to institutional control, both fiscal and substantive. Research universities have policies that are intended to guarantee reasonably free inquiry by faculty, but research topics usually are practically limited to those that find external financial support. Choice of research topics within institutions is also constrained by student and research peer pressure, and sometimes by the preferences of concerned publics, such as alumni and other donors, and, in public universities, by legislative preferences.

Research funds are distributed to institutions in two major ways: by formula and through competitive grants to individuals and groups. Each has intended outcomes that are sought by the sources of the funds. Formula, or directed, funding assumes that certain institutions and problems are sufficiently important, politically, economically, and socially, to be funded without competition among scientists or institutions on the merits of the specific research to be supported. For example, formula funds are distributed to colleges of agriculture by the USDA to ensure a strong agricultural research presence wherever agriculture is important. Competitive grants rely on competition among and judgment by scientists to distribute research support efficiently. The National Science Foundation distributes most of its research support this way, as do numerous other federally funded granting programs. Competition is intended to ensure quality and sharpen effort. In practice, both methods have strengths and weaknesses. It is often charged that formula funds are not directed to the highest scientific priorities and are used to support research by less-talented (and therefore less-competitive) researchers. On the other hand, politicians and practitioners often criticize competitive grants as not directed to the highest societal priorities and as a support system for scientists not sensitive to public needs. For many years, a prominent U.S. senator gave "Golden Fleece" awards to federally funded research projects that he and his staff thought worthy of ridicule on those grounds.

Most research administrators would agree that competitive funding of research has many advantages but that some directed funding is important. Directed funding can support important but "out-of-fashion" research and researchers, for example, that may be the source of future major advances. Less glamorous and scientifically interesting applied research and development that are nevertheless critical to using basic information for societal goals are often funded in this way.

THE REWARD SYSTEM

Research has shown that scientists regard recognition by their peers as their highest reward. The rewards that researchers seek, however, include a variety of other things. Salary, access to research funds, institutional prestige, appointment to membership in the National Academy of Science, and, on the down side, freedom from charges of fraud and duplicity all play a role in rewarding and motivating scientists. At universities, the achievement of a tenured appointment (essentially a promise of lifetime employment) can serve as a main motivator and reward. At research universities, research output is the main determinant of whether tenure is offered. Although this is often criticized on the grounds that once tenure is achieved effort may slacken, what evidence there is suggests that publication rate, an indicator of research output, does not decrease with the award of tenure and that those who publish early in their career continue to publish later.

SCIENCE AND THE POLICY PROCESS

Policy makers often wish to base policy on scientific findings. This is seen as a way to make policy "fact based" and therefore credible. Even though under a democratic form of government there is no particular reason why laws and regulations should be based on anything other than the will of the people, agreement on fact often precedes and facilitates agreement on a course of action. Thus, extensive analyses of ways to use science as policy input have been made. The complexity and contentious nature of many environmental and natural resource policy issues both call for resolution by appeal to science and make the appeal particularly difficult. Science-based assessments, which are attempts to answer questions from outside science (that is, from politicians and other policy makers) using data, people, and methods from science, are described as one method by which science can directly aid policy making. Others include testimony by scientists before policy-making bodies, such as congressional committees, the use of government and other scientists as expert witnesses in policy and legal disputes, and the preparation of science synthesis papers on major issues, such as the reports issued by the National Research Council.

Some research efforts are intended from the outset to contribute directly to the making of policy. At other times, science-based assessments are done to focus theory, data, and concepts from science on major questions. The National Acid Precipitation Assessment Program (NAPAP) was a pathbreaking effort to put science at the service of those making laws governing air pollution. It lasted ten years and cost more than a billion dollars in today's terms. It included synthesis of existing science, but it also funded a large body of additional research on different aspects of acid rain. Much was learned about acid rain and much was learned about the process of science-based assessment. Of the latter, perhaps the most important lessons were:

- The initial formulation of the policy questions to be answered is a major determinant of ultimate success. Both policy and science participants must agree on the questions, in detail, and on the boundary conditions for acceptable answers. For example, are answers to be framed as alternative courses of action with predicted consequences for each, or as a single, recommended course of action? What level of specificity is required for the answers to be politically useful?

- The timing and form of the communication of the results of the assessment are critical to their use, since policy will be made with or without the science input when the political time is right. If the results are not available at that time, or if they are not in readily understandable form, they will remain unused.

- The scientific participants must stay focused on the assessment question until results are successfully transmitted, no matter whether other interesting avenues of investigation are discovered in the process. There is a tendency (understandable but fatal to assessment) for scientists to substitute their own, inside science questions for the policy questions as the assessment evolves.

- Policy (political) commitment to the assessment process, including financial support and willingness to receive and consider results, must be maintained for the life of the assessment and its implementation.

Planning assessments is similar to planning original research in most respects; however, the added dimension of understanding and meshing with political processes makes many scientists nervous. The assessment process will likely be refined and more widely used in the future. Thus, the science community should participate fully in the development of the process for it to be seen as a willing and effective participant.

Science exists for two sometimes contradictory ends: to seek knowledge about the universe no matter what the consequences or who cares, and to help the larger world solve its problems and seize its opportunities. To be supported in the first, it is pretty generally agreed, science needs to deliver on the second. Effective research planning, and the creation and use of written research plans, is one key to success in both.

REFERENCES

Ackoff, R. L. 1962. *Scientific Method*. New York: Wiley.

Black, M. 1977. The objectivity of science. *Bulletin of the Atomic Scientists* (February): 55–60.

Bormann, B. T., P. G. Cunningham, M. H. Brookes, V. W. Manning, and M. W. Collopy. 1994. *Adaptive Ecosystem Management in the Pacific Northwest*. Gen. Tech. Rep. PNW-GTR-341. Portland, Ore.: U.S. Department of Agriculture, Forest Service, Pacific Northwest Research Station.

Bormann, B. T., P. G. Cunningham, and J. C. Gordon. 1996. *Best Management Practices, Adaptive Management, or Both?* Proc. Nat. SAF Conv. Portland, Maine, October 28–November 1, 1995.

Boulding, K. E. 1980. Science: Our common heritage. *Science* 207(22):831–837.

Bragg, L. 1966. The art of talking about science. *Science* 154:1613–1616.

Bush, V. [1945] 1990. *Science: The Endless Frontier. A Report to the President on a Program for Postwar Scientific Research*. Washington, D.C.: Republished by the National Science Foundation.

Caws, P. 1969. The structure of discovery. *Science* 166:1375–1380.

Chamberlain, T. C. 1965 [1899]. The method of multiple working hypotheses. *Science* 148:754–759.

Cochrane, W. G., and G. Cox. 1957. *Experimental Designs,* 2nd ed. New York: Wiley.

Cole, J., and H. Zuckerman. 1987. Marriage, motherhood and research performance in science. *Scientific American* 256(2):119–125.

REFERENCES

Council of Biological Editors. 1994. *Scientific Style and Format: The CBE Manual for Authors, Editors and Publishers.* Chicago: Council of Biological Editors.

Diamond, J. 1986. Overview: Laboratory experiments, field experiments and natural experiments. Pp. 3–22 in *Community Ecology,* J. Diamond and T. J. Case, eds. New York: Harper and Row.

Fisher, R. A. 1935. *The Design of Experiments.* Repr., New York: Hafner, 1960.

Ford, E. D. 2000. *Scientific Method for Ecological Research.* Cambridge: Cambridge University Press.

Gordon, J., and J. Berry. 1993. Chapter 1 in *Environmental Leadership: Developing Effective Skills and Styles,* J. Berry and J. Gordon, eds. Washington, D.C.: Island Press.

Gordon, J. C. 1999. History and assessments: Punctuated nonequilibrium. In *Bioregional Assessments Science at the Crossroads of Management and Policy,* K. N. Johnson et al., eds. Washington, D.C.: Island Press.

Hall, C. A. S., and J. W. Day. 1977. Systems and models: Terms and basic principles. In *Ecosystem Modeling in Theory and Practice,* C. A. S. Hall and J. W. Day, eds. New York: Wiley.

Heiligmann, R. B., and V. J. Rudolph. 1978. Guidelines for Program Participants. *Journal of Forestry,* March.

Hilborn, R. 1992. Can fisheries agencies learn from experience? *Fisheries* 17(4):6–14.

Holling, C. S., ed. 1978. *Adaptive Environmental Assessment and Management.* New York: John Wiley and Sons.

Holton, G. 1974. On being caught between dionysians and appollonians. *Daedalus* 103:65–80.

———. 1993. *Science and Anti-Science.* Cambridge, Mass: Harvard University Press.

Horgan, J. 1997. *The End of Science: Facing the Limits of Knowledge in the Twilight of the Scientific Age.* New York: Little, Brown and Co.

Illinois Institute of Technology, Research Institute. 1968. *Technology in Retrospect and Critical Events in Science.* Chicago: Illinois Institute of Technology.

Johnston, J. F. W. 1849. *Experimental Agriculture, Being the Results of Past and Suggestions for Future Experiments in Scientific and Practical Agriculture.* Report of the Rothamsted Experimental Station, U.K.

REFERENCES

Koshland, D. E. 1990. Two plus two equals five (Editorial). *Science* 247(4949):1381.

Kuhn, T. S. 1970. *The Structure of Scientific Revolutions,* 2nd ed. Chicago: University of Chicago Press.

Little, T. M., and F. J. Hills. 1978. *Agricultural Experimentation: Design and Analysis.* New York: Wiley.

Perry, C. R. 1993. The environment of words: A communications primer for leaders. In *Environmental Leadership: Developing Effective Skills and Styles,* J. Berry and J. Gordon, eds. Washington, D.C.: Island Press.

Platt, J. R. 1964. Strong inference. *Science* 146:347–353.

Popper, K. R. 1962. *Conjectures and Refutations: The Growth of Scientific Knowledge.* New York: Basic Books.

Quinn, J., and A. Dunham. 1983. On hypothesis testing in ecology and evolution. *American Naturalist* 122:602–617.

Ruby, L. 1950. *Logic: An Introduction.* Chicago: J. B. Lippincott.

Schrader-Frechette, K. S., and E. D. McCoy. 1993. *Method in Ecology Strategies for Conservation.* Cambridge: Cambridge University Press.

Shaw, B. T. 1967. *The Use of Quality and Quantity of Publication as Criteria for Evaluating Scientists.* Miscellaneous Publication No. 1041, Agricultural Research Service, USDA, Washington, D.C.

Snow, C. P. [1957] 1993. *The Two Cultures.* Cambridge: Cambridge University Press.

Sonnert, G. 1996. Gender equity in science: Still an elusive goal. *Science and Technology* 12:53–58.

Sonnert, G., and G. Holton. 1994. *Gender Differences in Science Careers: The Project Access Study.* New Brunswick, N.J.: Rutgers University Press.

———. 1996. Career patterns of women and men in the sciences. *American Scientist* 84:63–71.

Stoltenberg, C. H., K. D. Ware, R. J. Marty, R. D. Wray, and J. D. Wellons. 1970. *Planning Research for Resource Decisions.* Ames: Iowa State University Press.

Stock, M. 1985. *A Practical Guide to Graduate Research.* New York: McGraw-Hill.

Strauch, R. E. 1975. *"Squishy" Problems and Quantitative Methods.* Santa Monica, Calif.: The Rand Corporation P-5303.

Strunk, W., and E. B. White. 1979. *The Elements of Style.* 3rd ed. New York: MacMillan.

Thomas, J. W. 1993. Ethics for leaders. In *Environmental Leadership: Developing Effective Skills and Styles,* J. Berry and J. Gordon, eds. Washington, D.C., and Covelo, Calif.: Island Press.

Tufte, E. 1983. *The Visual Display of Quantitative Information.* Cheshire, Conn.: Graphics Press.

Waddington, C. H. 1977. *Tools for Thought: How to Understand and Apply the Latest Scientific Techniques of Problem Solving.* New York: Basic Books.

INDEX